Analysis of Singularities for Partial Differential Equations

Series in Applied and Computational Mathematics

ISSN: 2010-2739

Vol. 1: Analysis of Singularities for Partial Differential Equations
by Shuxing Chen (Fudan University, China)

Series in Applied and Computational Mathematics – Vol. 1

Analysis of Singularities for Partial Differential Equations

Shuxing Chen
Fudan University, China

NEW JERSEY • LONDON • SINGAPORE • BEIJING • SHANGHAI • HONG KONG • TAIPEI • CHENNAI

Published by

World Scientific Publishing Co. Pte. Ltd.

5 Toh Tuck Link, Singapore 596224

USA office: 27 Warren Street, Suite 401-402, Hackensack, NJ 07601

UK office: 57 Shelton Street, Covent Garden, London WC2H 9HE

British Library Cataloguing-in-Publication Data
A catalogue record for this book is available from the British Library.

ANALYSIS OF SINGULARITIES FOR PARTIAL DIFFERENTIAL EQUATIONS
Series in Applied and Computational Mathematics — Vol. 1

ISBN-13 978-981-4304-83-2
ISBN-10 981-4304-83-2

Printed in Singapore.

Preface

Regularity or singularity is a most important property of solutions to partial differential equations. In the study of the theory of partial differential equations and its applications people often confront various problems, which require to look for solutions. Besides the existence and uniqueness of solutions to these problems it is also essentially significant to master or understand various properties of solutions including regularity and singularity, periodicity, asymptotic behavior etc. Among these properties the regularity or singularity of the solutions often plays a crucial role or occupies in the core position due to the following reasons:

1. In the study of the modern theory of partial differential equations the concept of distribution is extensively applied. Correspondingly, the proof of the existence of solutions to various problems is often decomposed to two steps. The first step is to prove the existence of a distribution solution in a very weak sense, then the second step is to improve the regularity of the given solution, so that the weak solution will also be a solution in strong sense or even in classical sense. Obviously, the second step is nothing but the study of regularity of solutions.

2. Regularity and singularity are merely two sides of one matters. In many applications more attentions are often paid to the formation and the distribution of singularities of solutions. For instance, in the material sciences the singularity often corresponds to the crack of materials, in chemical reaction the formation of singularity often corresponds to explosion, and in gas dynamics the formation of singularity often corresponds to the formation of shock. In the problems on wave propagation the singularity of solutions can describe the wave front, so that the propagation of singularities can describe the physical phenomena of wave propagation precisely.

3. The regularity of a solution gives us the information whether the solution is continuous or smooth (having continuous derivatives). If the solution is continuous, then one can use the value of the solution at a given point to stand for its value in a neighborhood of this point. This gives us great convenience to understand the solution quantitatively, as well as to compute the solution numerically.

Further study also finds that the regularity of solutions is closely related to the existence and uniqueness of solutions. When various methods of functional analysis are applied in the study of partial differential equations one usually has to fix a functional space, to which the expected solution belongs. Such a space automatically implies a kind of regularity of the solution. Besides, the regularity of solutions is often expressed by some estimates, which are also helpful in the proof of existence and uniqueness of solutions.

Many textbooks and monographs on partial differential equations contain some discussions on regularity and singularity of solutions, but so far there is not any book especially expounding such a topic. Our book is named as *Analysis of singularities for partial differential equations*, because the crucial points, as well as the main difficulties, in the study of partial differential equations, are often companied by the formation, development and propagation of singularities. In this book we are trying to emphasize the importance of the study of this topic with offering a systematical results and methods on some typical problems. However, due to the rapid development of the study in this area, we can only give a general survey on the study of other problems. We hope both the detailed analysis on typical problems and the survey on other problems are helpful for reader's further research.

The writing of the book is partially supported by National Natural Science Foundation of China, the Key Grant of National Basic Research Program of China and the Doctoral Foundation of National Educational Ministry.

Shuxing CHEN
Fudan University, Shanghai, China

Contents

Chapter 1

Introduction to problems on singularity analysis

1.1 The classical singularity propagation theorem

In this section we first discuss the classical theorem of singularity propagation. Let us start with some examples. Consider the linear partial differential equation of first order in the space (x, y):

$$a(x,y)\frac{\partial u}{\partial x} + b(x,y)\frac{\partial u}{\partial y} = c(x,y), \tag{1.1}$$

where a, b, c are C^∞ function, and $a^2 + b^2 \neq 0$. Equation (1.1) can be solved by using the method of characteristics. The characteristics of Eq. (1.1) is the solution of the following system:

$$\frac{dx}{ds} = a(x,y), \quad \frac{dy}{ds} = b(x,y). \tag{1.2}$$

Suppose that there is a curve:

$$\ell: \ x = \xi(t), \ y = \eta(t), \ u = \zeta(t), \tag{1.3}$$

where $\xi(t), \eta(t), \zeta(t)$ are continuously differentiable, and $\xi_t^2 + \eta_t^2 \neq 0$. The problem to look for a function $u(x, y)$ satisfying Eq. (1.1) and $u(\xi(t), \eta(t)) = \zeta(t)$ is called Cauchy problem. To solve it one has to first solve the system of ordinary differential equations (1.2) with the initial data

$$x(0) = \xi(t), \quad y(0) = \eta(t). \tag{1.4}$$

The solutions of the above initial boundary value problems are a family of curves with one parameter t:

$$x = \xi(s,t), \quad y = \eta(s,t). \tag{1.5}$$

The family of integral curves covers a neighborhood Ω of $x = \xi(t),\ y = \eta(t)$. If $\partial(x, y)/\partial(s, t) \neq 0$ at any point in Ω, then there is a homeomorphism from (s, t) to (x, y). Integrating

$$\frac{du}{ds} = c(x(s,t), y(s,t)), \quad u|_{s=0} = \zeta(t)$$

gives the function $u(s,t)$. Then by using the inverse transformation of (1.5) we obtain the solution $u(x,y)$ of the problem (1.1), (1.2). Obviously, since the functions a, b only depend on x, y, then the procedure of solving the above two initial boundary value problems is equivalent to solve the following problem

$$\begin{cases} \dfrac{dx}{ds} = a, \quad \dfrac{dy}{ds} = b, \quad \dfrac{du}{ds} = c, \\ x|_{s=0} = \xi(t), \quad y|_{t=0} = \eta(t), \quad u|_{s=0} = \zeta(t). \end{cases} \tag{1.6}$$

The above argument is classical and can be found in [59].

Now let us analyze the singularity of the solution to the above problem. Singularity is the antonym of regularity. In the category of "C^∞", any point, where the solution is not C^∞, can be considered as singular point. Therefore, even for a continuously differentiable function $u(x,y)$ one can also discuss the set of its "singular" points. Suppose that the solution $u(x,y)$ of Eq. (1.1) is C^∞ at point (x_1,y_1), then one can draw a C^∞ curve ℓ : $x = \xi_1(t), y = \eta_1(t)$ through (x_1,y_1), such that the tangential direction of ℓ is transversal to the vector field everywhere. Denote $\zeta_1(t) = u(\xi_1(t), \eta_1(t))$, then the integral surface of Eq. (1.1) through $(\xi_1(t), \eta_1(t), \zeta_1(t))$ is nothing but $u = u(x,y)$. Obviously, the functions $x(s,t), y(s,t)$ obtained by using above-mentioned procedure are C^∞. Moreover, the Jacobian

$$\frac{\partial(x,y)}{\partial(s,t)} = a\eta_1'(t) - b\xi_1'(t) \neq 0$$

implies that $u(x,y)$ is C^∞ at any point along all characteristics through $(x_1, y_1, u(x_1,y_1))$. Conversely, if $u(x,y)$ is not C^∞ at (x_2,y_2), then $u(x,y)$ is also not C^∞ along the whole characteristics through (x_2,y_2) due to the uniqueness of the initial value problem of ordinary differential equations. In one word, *the C^∞ singularity of a differentiable solution propagates along characteristics of the equation.*

There are many different extension of the above conclusion. For instance, the corresponding conclusion for the piecewise smooth solutions is: *if a piecewise smooth solution of Eq. (1.1) has weak discontinuity (discontinuity of derivatives) on a curve, then the curve must be characteristics of the equation.*

Next let us consider the Cauchy problem of wave equation

$$\frac{\partial^2 u}{\partial t^2} - a^2 \frac{\partial^2 u}{\partial x^2} = 0 \tag{1.7}$$

satisfying the initial conditions

$$u(0,x) = 0, \quad u_t(0,x) = \phi(x), \tag{1.8}$$

where

$$\phi(x) = \begin{cases} (x^2-1)^2, & |x| \le 1; \\ 0, & |x| > 1. \end{cases} \tag{1.9}$$

Then the solution of Cauchy problem (1.8), (1.9) is

$$u(t,x) = \begin{cases} \frac{1}{2}[(x-at)^2-1]^2, & \text{if } |x-at| \le 1,\ |x+at| > 1; \\ \frac{1}{2}[(x+at)^2-1]^2, & \text{if } |x-at| > 1,\ |x+at| \le 1; \\ \frac{1}{2}[(x-at)^2-1]^2 + \frac{1}{2}[(x+at)^2-1]^2, & \\ \qquad \text{if } |x-at| \le 1,\ |x+at| \le 1; & \\ 0, \ \text{otherwise}. & \end{cases} \tag{1.10}$$

Obviously, the singularity of the solution $u(t,x)$ occurs on the characteristics through $(0, \pm 1)$. That is, *singularity propagates along characteristics.*

Is the conclusion still valid for more general partial differential equations? To explain it let us consider a linear partial differential equation of order m, which takes the form

$$\sum_{\alpha_1+\cdots+\alpha_n=\alpha,\ |\alpha)\le m} a_{\alpha_1,\cdots,\alpha_n}(x) \frac{\partial^\alpha u}{\partial x_1^{\alpha_1} \cdots \partial x_n^{\alpha_n}} = f(x_1,\cdots,x_n). \tag{1.11}$$

The weakly discontinuous solution is defined as

Definition 1.1. Assume that there is a C^∞ surface S in the domain Ω, where the function $u(x_1,\cdots,x_n)$ is defined. If $u \in C^\infty(\Omega \setminus S)$ is a solution of Eq. (1.11), $u \in C^{m-1}(\Omega)$ and the m-th derivatives of u has discontinuity of first class on S, then u is called **weakly discontinuous solution** of Eq. (1.11).

For the equation (1.11) the surface $\phi(x_1,\cdots,x_n) = 0$ is called its characteristic surface, if the equality

$$\sum_{\alpha_1+\cdots+\alpha_n=\alpha,\ |\alpha|=m} a_{\alpha_1,\cdots,\alpha_n}(x) \left(\frac{\partial \phi}{\partial x_1}\right)^{\alpha_1} \cdots \left(\frac{\partial \phi}{\partial x_n}\right)^{\alpha_n} = 0. \tag{1.12}$$

is satisfied at each point on $\phi = 0$.

Theorem 1.1. *If $u(x_1,\cdots,x_n)$ is a weakly discontinuous solution of Eq. (1.11), then the surface bearing the weak discontinuity of u must be a characteristic surface.*

Proof. Let P be any point on the surface S. Denote the equation of S near P by $\psi(x_1, \cdots, x_n) = 0$, which satisfies $\sum \psi_{x_i}^2 \neq 0$. Without loss of generality we assume that $\psi_{x_1} \neq 0$, then in the neighborhood of P the transformation

$$y_1 = \psi(x_1, \cdots, x_n),\ y_2 = x_2, \cdots,\ y_n = x_n \tag{1.13}$$

can be introduced. Obviously $\partial(y_1, \cdots, y_n)/\partial(x_1, \cdots, x_n) \neq 0$. Hence the transformation (1.13) is a homeomorphism. Under such a transformation the function $u(x)$ becomes $U(y) = u(x(y))$, and the surface S becomes $y_1 = 0$. Correspondingly, Eq. (1.11) is transformed into

$$\sum_{\alpha_1+\cdots+\alpha_n=\alpha,\ |\alpha|=m} a_{\alpha_1,\cdots,\alpha_n}(x(y)) \left(\frac{\partial \psi}{\partial x_1}\right)^{\alpha_1} \cdots \left(\frac{\partial \psi}{\partial x_n}\right)^{\alpha_n} \frac{\partial^m U}{\partial y_1^m}$$

$$+ \sum_{\beta_1+\cdots+\beta_n=\beta, |\beta|\le m, \beta_1<m} b_{\beta_1,\cdots,\beta_n}(y) \frac{\partial^\beta U}{\partial y_1^{\beta_1} \cdots \partial y_n^{\beta_n}} = f(x(y)). \tag{1.14}$$

According to the definition of weakly discontinuous solution all terms except the first one on the right hand side of Eq. (1.14) are continuous in the neighborhood of P. Taking the limit at the point P from both sides of S and then making subtraction of them one obtains

$$\sum_{|\alpha_1+\cdots+\alpha_n|=m} a_{\alpha_1,\cdots,\alpha_n}(x(y)) \left(\frac{\partial \psi}{\partial x_1}\right)^{\alpha_1} \cdots \left(\frac{\partial \psi}{\partial x_n}\right)^{\alpha_n} \left[\frac{\partial^m U}{\partial y_1^m}\right] = 0, \tag{1.15}$$

where $[\ \cdot\ ]$ means the jump of the quantity inside the bracket on S. Since derivatives of m-th order for $u(x)$ has discontinuity on S, then $\left[\frac{\partial^m U}{\partial y_1^m}\right] \neq 0$. Therefore, in order to let Eq. (1.15) hold, the coefficient of the term must be zero. This means that the surface $\psi = 0$ satisfies the requirement of being a characteristics. Hence S is characteristic surface. □

Remark 1.1. The above theorem asserts that the discontinuity of m-th derivatives of solutions to Eq. (1.11) is always distributed on the characteristic surface. The fact is also valid for weaker singularities. That is, if $u(x_1, \cdots, x_n)$ is a C^k solution of Eq. (1.11) with $k \ge m$, and the $(k+1)$-th derivatives has discontinuity of first class on S, then S must be characteristics of Eq. (1.11).

The conclusion given in Theorem 1.1 also holds for nonlinear equation. The general form of nonlinear partial differential equations of higher order is

$$F\left(x_1, \cdots, x_n, u, \cdots, \frac{\partial^{\alpha_1+\cdots+\alpha_n} u}{\partial x_1^{\alpha_1} \cdots \partial x_n^{\alpha_n}}, \cdots\right) = 0, \tag{1.16}$$

where F is a C^∞ function of its arguments $x_1, \cdots, x_n, u, \cdots, p_{\alpha_1, \cdots, \alpha_n}$. For a given solution $u(x)$, a surface $\psi(x_1, \cdots, x_n) = 0$ is called the characteristic surface, if the following equality

$$\sum_{|\alpha_1+\cdots+\alpha_n|=m} \frac{\partial F}{\partial p_{\alpha_1,\cdots,\alpha_n}} \left(x, u, \cdots, \frac{\partial^{\alpha_1+\cdots+\alpha_n} u}{\partial x_1^{\alpha_1} \cdots \partial x_n^{\alpha_n}}, \cdots\right) \times \left(\frac{\partial \psi}{\partial x_1}\right)^{\alpha_1} \cdots \left(\frac{\partial \psi}{\partial x_n}\right)^{\alpha_n} = 0, \tag{1.17}$$

holds on $\psi(x_1, \cdots, x_n) = 0$. In accordance, we have the following theorem:

Theorem 1.2. *If $u(x_1, \cdots, x_n)$ is the C^k solution of Eq. (1.16) with $k \geq m+1$, and its derivatives of $(k+1)$-th order have discontinuity of first class on S, then S must be the characteristic surface of Eq. (1.16).*

The proof of Theorem 1.2 is similar to that of Theorem 1.1. We leave it to readers.

Remark 1.2. In quasilinear or fully nonlinear case Since the characteristic surface S depends on the solution u, then the precise information on the singularity of solutions can only be obtained when the solution is obtained. This is different from the case of linear or semilinear equations. In the latter case the characteristics are independent of solutions.

Remark 1.3. For fully nonlinear partial differential equations (Eq. (1.16)), to determine the characteristics of Eq. (1.17) the derivatives of the coefficients of the latter will be used. Therefore, the solution u has to be C^{m+1} smooth.

Theorems 1.1 and 1.2 indicate that the singularity of weakly singular solutions must distribute on the characteristic surface. However, generally singularity may not appear on everywhere of a given characteristic surface. Then a question is how to give a more precise information on the distribution of singularities of solutions. Next we will discuss the problem for linear partial differential equations. The discussion is also essentially available to nonlinear equations.

Definition 1.2. Regarding Eq. (1.12) as a partial differential equation of first order, its characteristics is called **bicharacteristics** of the equation (1.11).

Denote Eq. (1.12) as $H\left(x_1, \cdots, x_n, \frac{\partial \phi}{\partial x_1}, \cdots, \frac{\partial \phi}{\partial x_n}\right) = 0$, it is a nonlinear partial differential equation of the function ϕ. Meanwhile, in the left hand side the function ϕ itself does not explicitly appear. For a given solution ϕ, the solution of the system of ordinary differential equations

$$\frac{dx_i}{ds} = \frac{\partial H}{\partial p_i} \tag{1.18}$$

satisfying the initial condition $x_i(0) = x_{i0}$ is called the characteristics of Eq. (1.12). Denote $p = \frac{\partial \phi}{\partial x_i}(x_1, \cdot, x_n)$ on the surface $\phi = 0$, we have

$$\frac{dp_i}{ds} = \sum_j \frac{\partial^2 \phi}{\partial x_i \partial x_j}\frac{dx_j}{ds} = \sum_j \frac{\partial^2 \phi}{\partial x_i \partial x_j} H_{p_j}.$$

Besides, differentiating Eq. (1.12) gives

$$H_{x_i} + \sum_j H_{p_j} \frac{\partial^2 \phi}{\partial x_i \partial x_j} = 0.$$

Hence $p_i(s)$ satisfies

$$\frac{dp_i}{ds} = -\frac{\partial H}{\partial x_i}(x, \phi_x). \tag{1.19}$$

Therefore, $(x_1(s), \cdots, x_n(s), p_1(s), \cdots, p_n(s))$ satisfies the system

$$\begin{cases} \dfrac{dx_i}{ds} = \dfrac{\partial H}{\partial p_i} \\ \dfrac{dp_i}{ds} = -\dfrac{\partial H}{\partial x_i} \end{cases} \quad (i = 1, \cdots, n), \tag{1.20}$$

where the arguments of H is $x_1, \cdots, x_n$ and $p_1, \cdots, p_n$. The solution of Eq. (1.20) is also called **bicharacteristic strip** of Eq. (1.11), while the by-characteristics of Eq. (1.11) is nothing but the project of the bicharacteristic strip.

Theorem 1.3. *The singularity of weakly discontinuous solutions to Eq. (1.11) on the characteristic surface $\phi = 0$ propagates along bicharacteristics.*

Proof. Similar to the proof of Theorem 1.1, we apply the transformation (1.13) to obtain

$$H\left(\frac{\partial \phi}{\partial x_1}\frac{\partial}{\partial y_1}, \frac{\partial \phi}{\partial x_2}\frac{\partial}{\partial y_1} + \frac{\partial}{\partial y_2}, \cdots, \frac{\partial \phi}{\partial x_n}\frac{\partial}{\partial y_1} + \frac{\partial}{\partial y_n}\right) u(x(y)) + Ku(x(y)) \\ = f(x(y)), \tag{1.21}$$

where K is a differential operator with order lower than m. Expanding the polynomial H and noticing Eq. (1.12) satisfied by ϕ we have

$$\sum_{i=2}^{n} H_{p_i}\left(\frac{\partial \phi}{\partial x_1}, \cdots, \frac{\partial \phi}{\partial x_n}\right) \cdot \frac{\partial}{\partial y_i}\left(\frac{\partial^{m-1} u}{\partial y_1^{m-1}}\right) + b\frac{\partial^{m-1} u}{\partial y_1^{m-1}} + \cdots = f, \tag{1.22}$$

where all omitted terms may contain differential operators with respect to y_1 of order less than $m-2$. Notice that u has weak discontinuity on S and $\frac{\partial^m u}{\partial y_1^m}$ has jump on S, then by differentiating Eq. (1.22) with respect to y_1 on both sides of S we have

$$\sum_{i=2}^{n} H_{p_i} \cdot \frac{\partial}{\partial y_i}\left(\frac{\partial^m u}{\partial y_1^m}\right) + b_1 \frac{\partial^m u}{\partial y_1^m} + \cdots = f, \tag{1.23}$$

where all omitted terms may contain differential operators with respect to y_1 of order less than $m-1$. Therefore, denote by

$$w = \left(\frac{\partial^m u}{\partial y_1^m}\right)_+ - \left(\frac{\partial^m u}{\partial y_1^m}\right)_-$$

the jump of $\left(\frac{\partial^m u}{\partial y_1^m}\right)$ on S, then w satisfies

$$\sum_{i=1}^{n} H_{p_i} \frac{\partial}{\partial y_i} w + b_1 w = 0, \tag{1.24}$$

or

$$\frac{dw}{ds} + b_1 w = 0.$$

Therefore,

$$w = w_0 \cdot \exp\left(-\int_{s_0}^{s} b_1 ds\right), \tag{1.25}$$

where w_0 is the value of w at $s = s_0$. Equation (1.25) means that the weak continuity propagates along bicharacteristics, i.e. if the jump of $\partial^m u$ is not zero at a point on a bicharacteristics, then it will never vanish at any point on this bicharacteristics. □

Example 1.1. Next we take the wave equation as an example to give more explanation. The wave equation in three dimensional space is

$$\frac{\partial^2 u}{\partial t^2} - \frac{\partial^2 u}{\partial x_1^2} - \frac{\partial^2 u}{\partial x_2^2} - \frac{\partial^2 u}{\partial x_3^2} = 0. \tag{1.26}$$

It can be used to describe the motion of sound waves, electromagnetic waves or optical waves. If $\phi(t, x_1, x_2, x_3) = 0$ is its characteristic surface, then the function ϕ should satisfy

$$\phi_t^2 - \phi_{x_1}^2 - \phi_{x_2}^2 - \phi_{x_3}^2 = 0. \tag{1.27}$$

When $\phi_t \neq 0$, one can solve $\phi(t, x_1, x_2, x_3) = 0$ to obtain $t = \psi(x_1, x_2, x_3)$, where ψ satisfies

$$\psi_{x_1}^2 + \psi_{x_2}^2 + \psi_{x_3}^2 = 1. \tag{1.28}$$

If $\phi(t, x_1, x_2, x_3) = 0$ is a surface bearing the singularities of u, then for any fixed t, the equation gives the location of the surface in (x_1, x_2, x_3) space at time t. The characteristics of Eq. (1.28) can be obtained from the projection of the integral curves

$$\frac{dx_i}{ds} = 2p_i, \quad \frac{dp_i}{ds} = 0 \tag{1.29}$$

on the space (x_1, x_2, x_3). Arbitrarily taking a point $(x_{10}, x_{20}, x_{30}, p_{10}, p_{20}, p_{30})$, satisfying $\sum p_{i0}^2 = 1$, then the solution of Eq. (1.29) with the initial data

$$x_i(0) = x_{i0},\ p_i(0) = p_{i0}\ (i = 1, 2, 3) \tag{1.30}$$

is

$$x_i = p_{i0}s + x_{i0},\ p_i = p_{i0}. \tag{1.31}$$

It is a straight line with direction (p_{10}, p_{20}, p_{30}). Since $(\sum(x_i - x_{i0})^2)^{1/2} = s$, and (p_{10}, p_{20}, p_{30}) is the normal direction of the surface (1.28), then Theorem 1.3 indicates that the weak singularity of the solution to Eq. (1.26) propagates along the normal direction of the wave front $\psi(x_1, x_2, x_3) = t$, and the speed of propagation is a constant. Therefore, if the wave equation is applied to describe optical waves, then the equation $\psi(x_1, x_2, x_3) = t$ gives the motion of the front of optical waves, and Eq. (1.31) indicates that the ray of light propagates along straight line in homogeneous media.

Consider the motion of optical waves in inhomogeneous media. The equation Eq. (1.26) becomes

$$\frac{\partial^2 u}{\partial t^2} - \frac{1}{n^2}\left(\frac{\partial^2 u}{\partial x_1^2} + \frac{\partial^2 u}{\partial x_2^2} + \frac{\partial^2 u}{\partial x_3^2}\right) = 0, \tag{1.32}$$

where $n = n(x_1, x_2, x_3)$ is the index of refraction. Similar to the above calculations we see that the function ψ satisfies

$$\psi_{x_1}^2 + \psi_{x_2}^2 + \psi_{x_3}^2 = n(x_1, x_2, x_3)^2. \tag{1.33}$$

In accordance, its characteristics satisfies

$$\frac{dx_i}{ds} = 2p_i, \quad \frac{dp_i}{ds} = (n^2)_{x_i}. \tag{1.34}$$

Generally, p_i is not constant because $n(x_1, x_2, x_3)$ is not constant. Therefore, the characteristics are not straight lines any more. This means that the light does not propagate along straight lines in inhomogeneous media.

1.2 Towards to modern theory

Although the classical theory mentioned in Section 1.1 gives much information on singularity propagation, it is still not precise enough. Particularly, the theory could not give a satisfied answer to some questions. For instance, consider the Cauchy problem in two dimensional space:

$$\begin{cases} \dfrac{\partial^2 u}{\partial t^2} = \dfrac{\partial^2 u}{\partial x^2} + \dfrac{\partial^2 u}{\partial y^2}, \\ u(0,x,y) = \phi(x,y), \quad u_t(0,x,y) = 0. \end{cases} \tag{1.35}$$

When $\phi(x,y)$ is taken as $\delta(x,y)$, its solution is

$$u_\delta(t,x,y) = \begin{cases} \dfrac{1}{2\pi}\dfrac{\partial}{\partial t}\dfrac{1}{\sqrt{t^2-x^2-y^2}}, & x^2+y^2<t^2; \\ 0, \quad x^2+y^2 \geq t^2. \end{cases} \tag{1.36}$$

On the other hand, if $\phi(x,y)$ is taken as a Heaviside function $\theta(x)$, then the solution is

$$u_\theta(t,x,y) = \begin{cases} 1, & x \geq t; \\ \frac{1}{2}, & -t \leq x \leq t; \\ 0, & x < t. \end{cases} \tag{1.37}$$

From the expression of the solution we see that the singularity of $u_\delta(t,x,y)$ at the origin propagates in all directions, while the singularity of $u_\theta(t,x,y)$ only appear on two planes $x = \pm t$.

According to the conclusion of Theorem 1.3 singularity propagates along bicharacteristics. Since the bicharacteristics of Eq. (1.35) issuing from $(0, x_0, y_0)$ is

$$t = s, \; x = ps + x_0, \; y = qs + y_0, \tag{1.38}$$

where p, q are constants satisfying $p^2 + q^2 = 1$. All thses bicharacteristics form a characteristic cone $(x-x_0)^2 + (y-y_0)^2 = t^2$ with $(0, x_0, y_0)$ as its vertex. The set of singularities of $u_\delta(t,x,y)$ coincides with this characteristic cone. However, in the case $\phi(x,y) = \theta(x)$, the initial data are discontinuous at every point on y-axis. Since the set of all bicharacteristics issuing from any point on y-axis form a solid wedge with y-axis as its edge. Then according to Theorem 1.3 every point in the solid wedge is possible to carry singularity of $u_\theta(t,x,y)$. However, the actual singularities only appear on the surface of the wedge. Why is it? We certainly want to know the reason and hope to find a way to describe more precise informations on the distribution of singularities.

The answer to the above problem will be given in the next chapter of our book. In history to understand such a phenomenon and give a conceivable explanation many mathematicians did extensive study on the theory of wave propagation, reflection, refraction etc. In the last century much progress was made particularly by J.Hadamard, I.G.Petrovsky, L.Garding, P.D.Lax etc. Their contributions greatly improved the understanding on the properties of solutions to partial differential equations describing various type of wave motion. Next we briefly introduce the idea in P.D.Lax's work [83], because the idea played important role for the development of the study of singularity analysis passing from the classical theory to the modern theory.

Given an initial boundary value problem of the partial differential equation

$$Pu \equiv \sum_{|\alpha|\le m} a_\alpha(t,x)\partial_{x_1}^{\alpha_1}\partial_{x_2}^{\alpha_2}\cdots\partial_{x_n}^{\alpha_n} u = 0, \tag{1.39}$$

$$u(0,x) = \phi(x). \tag{1.40}$$

Since the propagation of singularities is essentially equivalent to the process of wave propagation, one can consider the propagation of all waves with high frequency rather than the propagation of one wave front. To this end we may assume that the initial datum $\phi(x)$ takes the form $e^{i\xi\ell(x)}\psi(x)$, or a series with the form

$$\phi(x) \sim e^{i\xi\ell(x)}\left(\psi_0 + \frac{\psi_1}{\xi} + \cdots\right), \tag{1.41}$$

where ξ is a large parameter. Then we consider the solution with the form $u(t,x) = e^{i\xi\ell(t,x)}v(t,x,\xi)$, where $v(t,x,\xi)$ also has an asymptotic expansion of power series of ξ with minus power exponent. That is

$$u(t,x) \sim e^{i\xi\ell(t,x)}\left(v_0 + \frac{v_1}{\xi} + \cdots\right), \tag{1.42}$$

Assume that the differential operator P can be written as

$$P = P_m + P_{m-1} + \cdots,$$

where P_j is a differential operator of order j and has $p_j(x,\xi)$ as its symbol. Substituting u with the form (1.42) into Eq. (1.39) and rearranging all terms according to the power exponent of ξ we have

$$p_m(t,x,\ell_t,\ell_x)v_0\xi^m + \left(\sum_{j=0}^{n} p_m^{(j)}(t,x,\ell_t,\ell_x)\partial_{x_j}v_0 \right.$$
$$\left. + p_{m-1}(t,x,\ell_t,\ell_x)v_0 + p_m(t,x,\ell_t,\ell_x)v_1\right)\xi^{m-1} + \cdots, \tag{1.43}$$

where $p_m^{(j)}$ means the derivative of the polynomial $p_m(t, x, \zeta_0, \cdots, \zeta_n)$ with respect to ζ_j. Let the sum of all terms be zero, we see that $\ell(t, x)$ must satisfies

$$p_m(t, x, \ell_t, \ell_x) = 0, \tag{1.44}$$

which is called **eikonal** equation. Correspondingly, $v_k(t, s, \xi)$ should satisfy

$$\sum_{j=0}^{n} p_m^{(j)}(t, x, \ell_t, \ell_x)\partial_{x_j} v_0 + p_{m-1}(t, x, \ell_t, \ell_x) v_0 = 0,$$

$$\cdots\cdots$$

$$\left(\sum_{j=0}^{n} p_m^{(j)}(t, x, \ell_t, \ell_x)\partial_{x_j} + p_{m-1}(t, x, \ell_t, \ell_x)\right) v_k$$
$$+F_k(v_0, \cdots, v_{k-1}) = 0. \tag{1.45}$$

These equations are called **transport equations**, where F_k is C^∞ function of its arguments. We emphasize that the functions $v_0, \cdots, v_{k-1}$ in the expression of the equality to determine v_k are all given by previous equations, and the initial data of $\ell(t, x)$ and $v_k(t, x)$ can be determined by Eq. (1.41), i.e.

$$\ell(0, x) = \ell(x), \quad v_k(0, x) = \psi_k(x) \quad (k = 1, \cdots, n, \cdots). \tag{1.46}$$

By solving the Cauchy problems (1.44), (1.45), (1.46) the value of $v_k(t, x)$ can be obtained successively. If one does not care the precise meaning of the sum of the series (1.42), we have obtained the oscillatory solution of (1.45), (1.46) with high frequency.

We notice that once the solution $\ell(t, x)$ of Eq. (1.44) obtained, then for any constant c the equation $\ell(t, x) = c$ gives the characteristic surface of Eq. (1.39), whose section on the initial plane is $\ell(x) = c$. On the other hand in the equation of v_k, the functions $v_1, \cdots, v_{k-1}$ can be regarded as known. The differential operator of first order acting on each v_k is the same for different k. The direction of the operator is

$$\left(\frac{\partial p_m}{\partial \zeta_0}, \cdots, \frac{\partial p_m}{\partial \zeta_n}\right). \tag{1.47}$$

Obviously, the direction coincides with the direction of the bicharacteristics of the operator P. Hence each equation in Eq. (1.45) is an equation with differentiation along the bicharacteristics of P, so that they can be solved by using the characteristics method successively. Moreover, if the support of the initial datum $\phi(x)$ locates in a neighborhood of the point Q, then the support of all $v_k(t,x)$ will also locate in a neighborhood of the bicharacteristics γ issuing from Q. The fact simply means that the oscillation with high frequency propagates along bicharacteristics γ, like the propagation of singularities mentioned in Section 1.1.

We emphasize that the perturbation of the solution to Eq. (1.39) at the point Q will not propagate along all possible bicharacteristics issuing from Q. Indeed, the bicharacteristics γ propagating the perturbation depends on the value of $(\ell_t, \ell_{x_1}, \cdots, \ell_{x_n})$ on $t=0$, where $(\ell_{x_1}, \cdots, \ell_{x_n})$ on $t=0$ equals $(\psi_{x_1}, \cdots, \psi_{x_n})$, and the value of ℓ_t is determined by Eq. (1.44). Notice that Eq. (1.44) is a homogeneous algebraic equation of degree n for the variables $(\ell_t, \ell_{x_1}, \cdots, \ell_{x_n})$, which has n roots. Particularly, when Eq. (1.39) is hyperbolic with respect to t, these roots are all real. Therefore, we can have n real bicharacteristics issuing from the point Q. The oscillation with high frequency only propagates along these n bicharacteristics.

The above discussion indicates two facts. First, for the solutions of partial differential equations the oscillation with high frequency propagates along bicharacteristics like singularity does. Second, the path of propagation of oscillation with high frequency depends not only on the location where the oscillation occurs, but also on the direction of the oscillation. The facts give us an enlightenment that the singularity of solutions should be described by using location and direction as two basic elements, and the path of propagation depend on both. Indeed, this is an important idea for singularity analysis in modern theory of partial differential equations. Based on this idea the question raised in Section 1.1 can also be perfectly answered. In the next chapter we will indicate how to describe singularities of a given function by using these two elements, as well as how do they describe the propagation of singularities of solutions of partial differential equations.

Chapter 2

Singularity analysis for linear equations

2.1 Wave front set

As we indicated in the end of last section, location and direction are two main elements to describe the singularity of solutions to partial differential equations. These two elements can also offer more precise informations on the characters of propagation of singularities. To explain it we first study singularity of generalized functions in the sense of Schwartz's distribution. In the study of the theory of partial differential equations people also use the concept of generalized functions in other sense, like hyperfunctions in Sato's sense or generalized functions in Columbo's sense. Since those generalized functions are not as popular as Schwartz distributions, they will not be discussed in our book.

We notice that the singularity or regularity of a function is a local property. It means that in order to discuss the regularity of a given function one can consider its regularity in the neighborhood of each point in this domain. If a function u is infinitely differentiable, then one says that it is C^∞ smooth. Otherwise, one says that *it has singularity in C^∞ sense*, or simply *it has singularity*. The point is called the **singular point** of the function u. The closure of the set of all singular points of u is called **singular support** and denoted by singsupp u. However, for the functions with singularity at a same point, their property at this point can be quite different. Like in the examples of Section 1.2, both the function $\delta(x, y)$ and $\theta(x)$ have singularity at the origin, but the singularity of solution $u_\delta(t, x, y)$ propagates along all bicharacteristics issuing from the origin, while the singularity of solution $u_\theta(t, x, y)$ only propagates along the bicharacteristics locating on $x = \pm t$.

How to clearly describe the difference of the singularity for $\delta(x, y)$ and

$\theta(x)$ at the origin? A powerful tool is Fourier transformation, which connects the regularity of a function with the decay rate at infinity of its Fourier transformation. In the theory of Fourier transformation the Paley-Winner theorem confirms: denote by $\hat{u}(\xi)$ the Fourier transformation of a C_0^∞ function $u(x)$, for any integer N one can find a constant C_N such that

$$|\hat{u}(\xi)| \leq C_N(1+|\xi|)^{-N}, \quad \text{for any } N, \tag{2.1}$$

or

$$|\hat{u}(\xi)| = O(|\xi|)^{-N}, \quad \text{when } N \to \infty. \tag{2.2}$$

Moreover, the inverse of this proposition is also true. Therefore, if $u(x)$ is C^∞ at some point, then one can find a sufficiently small neighborhood of the point and a C_0^∞ function $\phi(x)$ supported in this neighborhood, such that the Fourier transformation $\widehat{\phi u}(\xi)$ satisfies Eq. (2.2). Conversely, if u is not C^∞ smooth at some point x, then for any C_0^∞ function, which is not zero at this point, the Fourier transformation $\widehat{\phi u}(\xi)$ could not have the estimate Eq. (2.2).

An important fact is that Eq. (2.2) offers a possibility to give a more precise description of the singularity of a function. When Eq. (2.2) does not hold, it is still possible that $\hat{u}(\xi)$ is rapidly decreasing in some direction in ξ-space. For instance, consider the Fourier transformation of Heaviside function $\theta(x)$. By taking $\phi(x,y)$ with the form $\psi_1(x)\psi_2(y)$, where both $\psi_1(x)$ and $\psi_2(y)$ are C_0^∞ functions, then the Fourier transformation of $\theta(x)\psi_1(x)\psi_2(y)$ is

$$\begin{aligned} F[\theta\psi_1\psi_2] &= \int_{-\infty}^{\infty}\int_{-\infty}^{\infty} e^{-1(x\xi+y\eta)}(\theta(x)\psi_1(x)\psi_2(y))dxdy \\ &= \hat{\psi_2}(\eta)\int_0^\infty e^{-ix\xi}\psi_1(x)dx. \end{aligned} \tag{2.3}$$

Consider the property of the above Fourier transformation as $(\xi,\eta) \to \infty$. In the right-hand side of Eq. (2.3) $\hat{\psi_2}(\eta)$ is rapidly decreasing as $|\eta| \to \infty$, while the another factor $\displaystyle\int_0^\infty e^{-ix\xi}\psi_1(x)dx$ is dominated by $\displaystyle\int_0^\infty |\psi_1(x)|dx$, so that is is a bounded quantity. Therefore, write $(\xi,\eta) = (\xi^2+\eta^2)^{1/2}(\tau_1,\tau_2)$, we have $(\xi^2+\eta^2)^{1/2} \leq C|\eta|$ along the direction (τ_1,τ_2), provided $\tau_2 \neq 0$. Then in this case

$$F[\theta\psi_1\psi_2] = O(|\eta|^{-N}) = O((\xi^2+\eta^2)^{-N/2}). \tag{2.4}$$

It turns out that the Fourier transformation is possibly not rapidly decreasing only at the direction $\tau_2 = 0$. Hence, if we call the directions, where the

Fourier transformation is rapidly decreasing, as "good" directions, and call other directions as "bad" directions, then all directions (τ_1, τ_2) with $\tau_2 \neq 0$ are good, and only the direction $(1,0)$ is possibly bad (readers may prove themselves that the direction $(1,0)$ is actually bad.)

The situation is totally different for $\delta(x,y)$. If $\phi(0) \neq 0$, then

$$F[\phi\delta](\xi,\eta) = \langle \phi\delta, e^{-i(x\xi+y\eta)} \rangle = \phi(0) \neq 0. \tag{2.5}$$

Therefore, $\widehat{\phi\delta}(\xi,\eta)$ will never decrease at any direction as $(\xi,\eta) \to \infty$, i.e. any direction in (ξ,η) space is bad direction.

Based on the above analysis we introduce the concept of **wave front set** (denoted by $WF(u)$ or WFu) to describe the property of a function at its singular point. Before giving the definition of wave front set let us first give the concept of conical neighborhood. If $V \in R^n_\xi$ is a set satisfying the condition that $\xi \in V$ and $t > 0$ implies $tV \in V$, then the set is called **cone.** If V is an open set in R^n_ξ and $\xi_0 \in V$, then V is called **conical neighborhood** of ξ_0.

Definition 2.1. For a given function $u \in \mathscr{D}'(\Omega)$, its **wave front set** $WF(u)$ is defined as the following subset in $\Omega \times R^n_\xi \setminus O$, such that for any $(x_0,\xi_0) \notin WF(u)$, there is a neighborhood ω of x_0 and a conical neighborhood V of ξ_0, such that for any function $\phi(x) \in C_0^\infty(\omega)$ and any integer N one can find a constant C_N such that the inequality

$$|\widehat{\phi u}(\xi)| \leq C_N(1+|\xi|)^{-N}, \quad \forall \xi \in V \tag{2.6}$$

holds.

The next theorem gives the relation between the singular support and the wave front set for a given function.

Theorem 2.1. *Let $x_0 \in \Omega$ is any given point, if $(x_0,\xi) \in WF(u)$ for any $\xi \in R^n \setminus 0$, then $x_0 \in$* singsupp *u.*

Proof. According to the definition of the wave front set, for any ξ one can find a neighborhood ω_ξ of x_0 and a conical neighborhood V_ξ of ξ, such that Eq. (2.6) holds for any $C_c^\infty(\omega_\xi)$ function $\phi_\xi(x)$. Since the unit sphere in R^n_ξ is a compact set, then one can find finite V_j, such that $\cup_j V_j = R^n_\xi \setminus 0$. Let $\omega = \cap_j \omega_j$, and take $\phi \in C_c^\infty(\omega)$, then $\widehat{\phi u}(\xi)$ is rapidly decreasing on any V_j as $|\xi| \to \infty$. Hence for any integer N, there is C_N, such that

$$|\widehat{\phi u}(\xi)| \leq C_N(1+|\xi|)^{-N}, \quad \forall \xi \in R^n \tag{2.7}$$

holds. Then the Paley-Winner theorem confirms $\phi u \in C_0^\infty(\omega)$. In view of $\phi(x_0) \neq 0$ we know u is C^∞ at x_0. □

Theorem 2.2. *Denote by* Π_x *the projection* $(x,\xi) \mapsto x$, *then*

$$\Pi_x WF(u) = \text{singsupp } u. \tag{2.8}$$

Proof. If $x_0 \notin \text{singsupp} u$, then u is C^∞ in a neighborhood ω of x_0. Then for any $\phi(x) \in C_c^\infty(\omega)$, $\widehat{\phi u}$ is rapidly decreasing in any direction. This means that for any ξ the point $(x_0, \xi) \notin WF(u)$. Hence $x_0 \notin \Pi_x WF(u)$.□

These two theorems clearly indicate that the wave front set is a finer description than singular support for a distribution.

The wave front set for some simple functions can be obtained by direct computation. For instance

$$WF(\delta) = \{(0,0,\xi,\eta);\ \forall\ (\xi,\eta) \in R^2\},$$

$$WF(\theta(x)) = \{(0,y,\xi,0)\}.$$

For more general functions people have to use some operation rule to obtain their wave front sets. For instance, if u and v are two functions defined on same open set, then we obviously have $WF(u+v) \subset WF(u) \cup WF(v)$. However for product of functions the corresponding rule is not so simple. In this case we have the following theorem.

Theorem 2.3. *If* u *is a distribution defined on* Ω, $a \in C^\infty(\Omega)$, *then*

$$WF(au) \subset WF(u). \tag{2.9}$$

Proof. If $(x_0, \xi_0) \notin WF(u)$, then according to Definition 2.1 one can find a neighborhood ω of x_0 and a conical neighborhood V of ξ_0 such that for any $\phi(x) \in C_0^\infty(\omega)$ Eq. (2.6) holds. Next we assume that a is compactly supported, otherwise a can be replaced by ζa, where ζ is a C_0^∞ function, which equals 1 in ω. From the formula of Fourier transformation of the product of two functions we have

$$\widehat{\phi a u}(\xi) = \int \hat{a}(\xi - \eta)(\widehat{\phi u})(\eta) d\eta. \tag{2.10}$$

Taking V_1 be another conical neighborhood of ξ_0, such that $V_1 \subset\subset V$ (the closure of V_1 in $R^n \setminus 0$ is included in V). For $\xi \in V$, Eq. (2.10) can be written as

$$\widehat{\phi a u}(\xi) = \int_V \hat{a}(\xi - \eta)(\widehat{\phi u})(\eta) d\eta + \int_{R^n \setminus V} \hat{a}(\xi - \eta)(\widehat{\phi u})(\eta) d\eta.$$

For the first term in the right hand side, since $\hat{a}$ is rapidly decreasing, $\widehat{\phi u}$ is rapidly decreasing in V, then for any N_1, by taking $N = N_1 + n + 1$ we have

$$\begin{aligned}
&(1+|\xi|)^{N_1}\left|\int_V \hat{a}(\xi-\eta)\widehat{\phi u}(\eta)d\eta\right| \\
&\leq \int_V (1+|\xi-\eta|)^{N_1}(1+|\eta|)^{N_1}|\hat{a}(\xi-\eta)|\cdot|\widehat{\phi u}(\eta)|d\eta \\
&\leq C\int_V (1+|\eta|)^{-n-1}d\eta \;\leq\; C'.
\end{aligned}$$

For the second term, since $V_1 \subset\subset V$, $\xi \in V_1$, $\eta \notin V$, then there is a constant $c > 0$, such that $|\xi - \eta| \geq c(|\xi| + |\eta|)$. Then for any N_1, by taking $N = \max(N_1, n+1)$, we have

$$\begin{aligned}
&(1+|\xi|)^{N_1}\left|\int_{R^n\setminus V} \hat{a}(\xi-\eta)\widehat{\phi u}(\eta)d\eta\right| \\
&\leq (1+|\xi|)^{N_1}\int_{R^n\setminus V} (1+|\xi|+|\eta|)^{-N}(1+|\eta|)^{-N}|d\eta \\
&\leq C\int_{R^n\setminus V} (1+|\eta|)^{-n-1}d\eta \;\leq\; C'.
\end{aligned}$$

Therefore, for any $\xi \in V_1$, $(1+|\xi|)^{-N_1}|\widehat{\phi a u}(\xi)|$ is bounded for any N_1. Hence $(x_0, \xi_0) \notin WF(au)$. □

Remark 2.1. From the proof of the above theorem we obtain an equivalent definition of the wave front wet as follows.

Definition 2.2. For a given function $u \in \mathscr{D}'(\Omega)$, its **wave front set** $WF(u)$ is defined as the following subset in $\Omega \times R^n_\xi \setminus O$, such that for any $(x_0, \xi_0) \notin WF(u)$ there is a neighborhood ω of x_0, a conical neighborhood V of ξ_0 and a function $\phi(x) \in C_0^\infty(\omega)$ satisfying $\phi(x_0) \neq 0$, such that for any integer N one can find a constant C_N, which let the following inequality

$$|\widehat{\phi u}(\xi)| \leq C_N(1+|\xi|)^{-N}, \quad \forall \xi \in V \tag{2.11}$$

hold.

Apparently, Definition 2.1 is more restrict than Definition 2.2, because the estimate Eq. (2.6) should be valid for **all** $\phi \in C_0^\infty(\omega)$, while the estimate in Definition 2.2 is valid only for **one** $C_0^\infty(\omega)$ function. However, these two definitions are essentially equivalent. Indeed, if $(x_0, \xi_0) \notin WF(u)$ in the sense of Definition 2.2, then we can take another neighborhood ω_1 of x_0, such that $\phi(x) \neq 0$ at every point of ω_1. Then for any function $\phi_1(x) \in C_0^\infty(\omega_1)$, let

$$\phi_2(x) = \frac{\phi_1(x)}{\phi(x)} \in C_0^\infty(\omega_1),$$

we have $\phi_1 u = \phi_2 \cdot (\phi u)$. According to Theorem 2.3 the rapidly decreasing of $\widehat{\phi u}$ in some open conical set implies the rapid decreasing of $\widehat{\phi_1 u}$ in this conical set. Hence $(x_0, \xi_0) \notin WF(u)$ in the sense of Definition 2.1 also holds.

Theorem 2.4. *If $(WF(u) + WF(v)) \cap O_x = \emptyset$, then the product of u and v is well-defined, and*

$$WF(uv) \subset (WF(u) + WF(v)) \cup WF(u) \cup WF(v), \tag{2.12}$$

where $WF(u) + WF(v)$ is the set

$$\{(x, \xi); \xi = \xi_1 + \xi_2, (x, \xi_1) \in WF(u), (x, \xi_2) \in WF(v)\},$$

O_x is $\{(x, 0); x \in O\}$.

Proof. The main idea is to make the convolution of the Fourier transformation of u and v, and then define the product of u and v by the inverse Fourier transformation of the convolution. Here the assumptions ensure the existence of the convolution.

Let us proceed our discussion in a neighborhood of $x \in \Omega$. For any fixed s, take a conical neighborhood V_1 of $WF(u)$ and a conical neighborhood V_2 of $WF(v)$, then $(WF(u) + WF(v)) \cap O_x = \emptyset$ implies that $(WF(u) + WF(v))$ is also a closed cone (except at the origin). Let V_3 be a conical neighborhood containing $(WF(u) + WF(v))$. For $i = 1, 2, 3$, define V_i' be the neighborhood of $\bar{V}_i$ and then consider the behavior of the Fourier transformation of uv outside $V_1' \cup V_2' \cup V_3'$. Here we assume that V_i, V_i' $(i = 1, 2, 3)$ are sufficiently close to $WF(u), WF(v), WF(u) + WF(v)$ respectively, and $V_1' + V_2' \subset V_3$ holds.

Now take ϕ is a C_0^∞ function, which does not vanish at x, which let $\widehat{\phi u}$ is rapidly decreasing outside V_1 and $\widehat{\phi v}$ is rapidly decreasing outside V_2, then for any $\xi \notin \cup V_j'$ and any integer N,

$$|(1 + |\xi|)^N F(\phi^2 uv)(\xi)| = \left| \int (1 + |\xi|)^N \widehat{\phi u}(\xi - \eta) \widehat{\phi v}(\eta) d\eta \right|.$$

Next we estimate the value of the integral in the domains

$$\left\{ |\xi - \eta| \leq \frac{|\xi|}{2}, \ \eta \notin V_2 \right\}, \left\{ |\xi - \eta| \leq \frac{|\xi|}{2}, \ \eta \in V_2 \right\},$$

$$\left\{|\xi-\eta|>\frac{|\xi|}{2},\ \eta\notin V_2\right\},\left\{|\xi-\eta|>\frac{|\xi|}{2},\ \eta\in V_2\right\}$$

respectively. Since ϕ is compactly supported, then there is N_0, such that

$$|\widehat{\phi u}(\xi-\eta)|\le C(1+|\xi-\eta|)^{N_0},\quad |\widehat{\phi v}(\eta)|\le C(1+|\eta|)^{N_0}.$$

Let M be a sufficiently large number, then in $\left\{|\xi-\eta|<\frac{|\xi|}{2},\eta\in V_2\right\}$

$$|\widehat{\phi v}(\eta)|\le C_M(1+|\eta|)^{-M}\le C'_M(1+|\xi|)^{-M},$$

$$|\widehat{\phi u}(\xi-\eta)|\le C(1+|\xi|)^{N_0}(1+|\eta|)^{N_0}.$$

By taking M large enough, we see that

$$I=\int_{|\xi-\eta|\le\frac{|\xi|}{2},\ \eta\notin V_2}(1+|\xi|)^N\widehat{\phi u}(\xi-\eta)\widehat{\phi v}(\eta)d\eta$$

is rapidly decreasing.

Consider the integral in the domain $\{|\xi-\eta|\le\frac{|\xi|}{2},\ \eta\in V_2\}$. In this region $\xi-\eta\notin V_1$, because otherwise $\xi=\xi-\eta+\eta\subset V_1+V_2\subset V'_3$ will lead to contradiction. Besides, the fact that $\eta\in V_2,\xi\notin V'_2$ gives $|\xi-\eta|>c(|\xi|+|\eta|)$. In view of that $\widehat{\phi u}$ is rapidly decreasing outside V_1, we have

$$|\widehat{\phi u}(\xi-\eta)|\le C_M(1+|\xi-\eta|)^{-M}\le C'_M(1+|\xi|)^{-M}(1+|\eta|)^{-M},$$

Hence

$$II=\int_{|\xi-\eta|\le\frac{|\xi|}{2},\ \eta\in V_2}(1+|\xi|)^N\widehat{\phi u}(\xi-\eta)\widehat{\phi v}(\eta)d\eta$$

is rapidly decreasing, because M can be arbitrarily large. Similarly, the integrals in the domains

$$\left\{|\xi-\eta|>\frac{|\xi|}{2},\ \eta\notin V_2\right\},\left\{|\xi-\eta|>\frac{|\xi|}{2},\ \eta\in V_2\right\}.$$

are also bounded.

Summing up, we have proved that if $\xi\notin V'_1\cup V'_2\cup V'_3$. then for any N,

$$F(\phi^2uv)(\xi)=O((1+|\xi|)^{-N}).\tag{2.13}$$

Notice that V'_1,V'_2,V'_3 can be arbitrarily close to $WF(u),WF(v),WF(u)+WF(v)$, then Eq. (2.12) is established. □

Assume that a homeomorphism $y = \psi(x)$ defined on the domain $\Omega \subset R^n_x$ transforms the domain to a domain G in R^n_y. Then a distribution u defined on G corresponds to an induced distribution $\psi^* u$ defined on Ω: $(\psi^* u)(x) = u(\psi(x))$. In this case the set of singular points of $\psi^* u$ is the inverse image of the set of singular points of u. A natural question is that whether it is possible to determine the wave front set of $\psi^* u$ by using $WF(u)$? The following theorem gives an answer.

Theorem 2.5. *Assume that ψ is the above C^∞ homeomorphism, then*

$$WF(\psi^* u) = \{(x, \xi); (\psi(x), ({}^t\psi')^{-1}\xi) \in WF(u)\}. \tag{2.14}$$

Proof. Consider the property of the following integral as $|\xi| \to \infty$:

$$\begin{aligned} I &= \int e^{-i\langle x,\xi\rangle}(\psi^* u)(x)\phi(x)dx \\ &= \int e^{-i\langle \psi^{-1}(y),\xi\rangle} u(y)\phi(\psi^{-1}(y))(\psi^{-1}(y))' dy. \end{aligned}$$

Denote $\phi_1(y) = \phi(\psi^{-1}(y))(\psi^{-1}(y))'$, and make a function $\phi_2(y) \in C_0^\infty(G)$, which is equal to 1 on the support of ϕ_1, then the above equality can be written as

$$\int e^{-i\langle \psi^{-1}(y),\xi\rangle} u(y)\phi_1(y)\phi_2(y)dy.$$

By using Fourier transformation we have

$$I = \frac{1}{2\pi}\iint e^{-i\langle \psi^{-1}(y),\xi\rangle - \langle y,\eta\rangle}\phi_1(y)dy\widehat{\phi_2 u}(\eta)d\eta. \tag{2.15}$$

Now if (x, ξ) satisfies $(\psi(x), ({}^t\psi')^{-1}\xi) \notin WF(u)$, then from the definition of the wave front set we can find a neighborhood ω of the point $\psi(x)$ and a conical neighborhood V of the direction $({}^t\psi')^{-1}\xi$, such that for any $y \in \omega$ and $({}^t\psi'(y))^{-1}\xi \in V' \subset\subset V$. Moreover, for any $\zeta(y)$ satisfying $\text{supp}\zeta \subset \omega$

$$|\widehat{\zeta u}(\eta)| \le C_N(1 + |\eta|)^{-N}, \quad \forall \eta \in V. \tag{2.16}$$

Obviously, the support of $\phi(x)$, $\phi_1(y)$ and $\phi_2(y)$ can be chosen as sufficiently small, then $\text{supp}\phi_2 \subset \omega$. Hence the inequality (2.16) still holds when ζ is replaced by $\phi_2(y)$.

Decompose the integral in Eq. (2.15) as the sum of the corresponding integrals on V and $R^n_\eta \setminus V$. For the integral on V, we make an operator L_1 as

$${}^tL_1 = i\sum_j \frac{\partial_{y_j}\langle \psi^{-1}(y),\xi\rangle}{|\nabla_y\langle \psi^{-1}(y),\xi\rangle|^2}\partial_{y_j},$$

where the denominator does not vanish for $\xi \neq 0$. L_1 satisfies

$$ {}^tL_1 e^{-i\langle\psi^{-1}(y),\xi\rangle} = e^{-i\langle\psi^{-1}(y),\xi\rangle}. $$

Now for any integer M, by taking $N = M + n + 1$ we have $|\widehat{\phi_2 u}(\eta)| \leq C_N(1+|\eta|)^{-N}$. Therefore,

$$ \begin{aligned} &\int_V \int_{R^n_y} e^{-i(\langle\psi^{-1}(y),\xi\rangle - \langle y,\eta\rangle)}\phi_1(y)dy\widehat{\phi_2 u}(\eta)d\eta \\ &\leq C\int_V \left| \int_{R^n_y} e^{-i\langle\psi^{-1}(y),\xi\rangle} L_1^M(e^{-i\langle y,\eta\rangle}\phi_1(y))dy \right| \cdot |\widehat{\phi_2 u}(\eta)|d\eta \\ &\leq \int_V (1+|\xi|)^{-M}(1+|\eta|)^M |\widehat{\phi_2 u}(\eta)|d\eta \\ &\leq C(1+|\xi|)^{-M}. \end{aligned} $$

It is then rapidly decreasing as $|\xi| \to \infty$.

Consider the integration on $R^n_\eta \setminus V$. In view of

$$ |\nabla_y(\langle\psi^{-1}(y),\xi\rangle - \langle y,\eta\rangle)| \geq c(|\xi| + |\eta|), $$

then we can define an operator

$$ {}^tL_2 = \sum_j \frac{i\partial_{y_j}(\langle\psi^{-1}(y),\xi\rangle - \langle y,\eta\rangle)}{|\nabla_y(\langle\psi^{-1}(y),\xi\rangle - \langle y,\eta\rangle)|^2}\partial_{y_j}, $$

which satisfies

$$ {}^tL_2 e^{-i(\langle\psi^{-1}(y),\xi\rangle - \langle y,\eta\rangle)} = e^{-i(\langle\psi^{-1}(y),\xi\rangle - \langle y,\eta\rangle)}. $$

Notice that $\widehat{\phi_2 u}(\eta)$ is slowly increasing as $|\eta| \to \infty$, then there is a number N_0, such that

$$ |\widehat{\phi_2 u}(\eta)| \leq C(1+|\eta|)^{-N_0}. $$

Hence by taking $N = M + N_0 + n + 1$ we have

$$ \begin{aligned} &\left| \int_{R^n_\eta \setminus V} \int_{R^n_y} e^{-i(\langle\psi^{-1}(y),\xi\rangle - \langle y,\eta\rangle)}(L_2)^N \phi_1(y)dy\widehat{\phi_2 u}(\eta)d\eta \right| \\ &\leq C\int_{R^n_\eta \setminus V} (1+|\xi|+|\eta|)^{-M-N_0-n-1}(1+|\eta|)^{N_0}d\eta \\ &\leq C(1+|\xi|)^{-M}, \end{aligned} $$

which is also rapidly decreasing as $|\xi| \to \infty$. Summing up we know the integral I is rapidly decreasing. Then the definition of the wave front set implies

$$ (x,\xi) \notin WF(\psi^* u), \tag{2.17} $$

which indicates

$$WF(\psi^* u) \subset \{(x,\xi); (\psi(x), ({}^t\psi')^{-1}\xi) \in WF(u)\}. \tag{2.18}$$

Obviously, the inverse inclusion in Eq. (2.18) is also valid, because ψ is a homeomorphism. Equation (2.14) is thus obtained. □

The above theorem indicates that in the process of coordinate transformation of variables the wave front set of a given function varies according the rule of coordinate transformation on cotangent bundle. Therefore, for a distribution defined on a differential manifold one can defined its wave front set on the cotangent bundle of the manifold. When the manifold locally maps to an open set Ω of a Euclid space, its wave front set maps to a conical subset of $T^*(\Omega)$.

In the study of the action of pseudodifferential operators on distributions we have the following theorems:

Theorem 2.6. *If A is a pseudodiferential operator with its symbol $a \in \mathscr{S}^\infty(\Omega)$, $u \in \mathscr{E}'(\Omega)$, then*

$$WF(Au) \subseteq WF(u), \quad WF(u) \subseteq WF(Au) \cup Char(A), \tag{2.19}$$

where $Char(A)$ stands for the characteristic set of A, i.e. the set of all (x,ξ) satisfying $a(x,\xi) = 0$.

Theorem 2.7. *Assume that $u(t,x) \in C^\infty((\alpha,\beta), \mathscr{E}'(\Omega_x))$, $A = a(x, D_x)$ is a pseudodifferential operator defined on Ω_x. If $(t_0, x_0; \tau_0, \xi_0) \notin WF(u)$, then $(t_0, x_0; \tau_0, \xi_0) \notin WF(Au)$.*

The proof of these two theorem can be find in [40],[74].

The wave front set defined in Definition 2.1 describes the singularity of a given function in C^∞ category. In some cases one needs to describe the singularity for a given function with finite order. For instance, for any given real number s, one can say "a distribution u is H^s smooth", which simply means $u \in H^s$. Furthermore, one can also say that u has singularity in the sense of H^s, if u is not in H^s. Corresponding wave front set $WF_s(u)$ of a distribution $u \in \mathscr{D}'(\Omega)$ can also be well-defined.

Definition 2.3. For $u \in \mathscr{D}'(\Omega)$, a point $(x_0, \xi_0) \notin WF_s(u)$ means that there is a neighborhood ω of x_0 and a conical neighborhood V of ξ_0, such that for any $\phi \in C_0^\infty(\omega)$

$$(1 + |\xi|^2)^{s/2} \widehat{\phi u} \in L^2(V). \tag{2.20}$$

By using the concept of the wave front set $WF(u)$ one can also define the microlocal Sobolev regularity for a distribution u. One says $u \in H^s(x_0, \xi_0)$ (or $u \in H^s_{(x_0,\xi_0)}$), if and only if $(x_0, \xi_0) \notin WF_s(u)$. Meanwhile, the property of pseudodifferential operators indicates that if A is a pseudodifferential operator of order m and $u \in H^s_{(x_0,\xi_0)}$, then $Au \in H^{s-m}_{(x_0,\xi_0)}$.

Obviously, the wave front set and the microlocal Sobolev regularity is two different descriptions of a same object. Moreover, we can also introduce a regularity function $s_u(x, \xi)$ to describe the regularity of a distribution (see [72]):

$$s_u(x, \xi) = \sup\{s;\ u \in H_s(x, \xi)\}. \tag{2.21}$$

Obviously, the regularity function is a lower semi-continuous function. For any fixed (x_0, ξ_0) and $\epsilon > 0$ there is a neighborhood ω of x_0 and a conical neighborhood V of ξ_0 such that for any $\phi(x) \in C_0^\infty(\omega)$

$$\widehat{\phi u}(\xi)(1 + |\xi|^2)^{(s-\epsilon)/2} \in L^2(V), \tag{2.22}$$

where $s = s(x_0, \xi_0)$. In Chapter 3 we will discuss the regularity function again and will apply it to treat problems on interaction of singularities.

2.2 Singularity propagation theorem for equations of principal type

Based on the new understanding on the singularities of functions we study the propagation of singularities of solutions to partial differential equations. We first study the case for linear equations, because in this case the characteristics is independent of solutions, so that the statement and the proof of the corresponding theorems will be simpler. In this section we restrict ourselves to the case, when the partial differential equation under consideration only has simple characteristics.

It is a main success in the theory of microlocal analysis, that one can use the wave front set to give a precise description of singularity propagation. The next theorem and its proof is taken from [108]. In Chapter 5 we will give a new proof on the theorem. Other proof can also be found in [71].

Denote $D_j = \frac{1}{i}\partial_{x_j}$, $D^\alpha = D_1^{\alpha_1} \cdots D_n^{\alpha_n}$. Let $P = p(x, D) = \sum_{|\alpha| \le m} a_\alpha D^\alpha$ be a given partial differential operator of m-th order with its principle symbol $p_m(x, \xi) = \sum_{|\alpha|=m} a_\alpha \xi^\alpha$ being real function. Then the vector field

$$H_{p_m} = \left\{ \frac{\partial p_m}{\partial \xi_1}, \cdots, \frac{\partial p_m}{\partial \xi_n}, -\frac{\partial p_m}{\partial x_1} \cdots, -\frac{\partial p_m}{\partial x_n} \right\} \tag{2.23}$$

is called **Hamilton vector field** of $p_m(x,\xi)$ (or operator P). The integral curve of the vector field is the solution of the following system

$$\begin{cases} \dfrac{dx_j}{ds} = \dfrac{\partial p_m}{\partial \xi_j} \\ \dfrac{d\xi_j}{ds} = -\dfrac{\partial p_m}{\partial x_j} \end{cases} \quad (j = 1, \cdots, n), \tag{2.24}$$

which is called **bicharacteristic strip**. The projection of the bicharacteristic strip on x space is called **bicharacteristics**. Obviously, if $p_m(x_0,\xi_0) = 0$, then $p_m(x(s),\xi(s)) \equiv 0$ due to

$$\frac{d}{ds} p_m(x(s),\xi(s)) = \sum_j \left(\frac{\partial p_m}{\partial x_j}\frac{dx_j}{ds} + \frac{\partial p_m}{\partial \xi_j}\frac{d\xi_j}{ds} \right) = 0. \tag{2.25}$$

Such a bicharacteristic strip is called **null bicharacteristic strip**.

The set of all (x,ξ) satisfying $p_m(x,\xi) = 0$ is called **characteristic set**. In this section we only study the operators satisfying the condition $\nabla_\xi p_m(x,\xi) \neq 0$ on its characteristic set. Such an operator is called **operator of principal type** (or **operator of principal type in strict sense**). The hyperbolic operator is an example of the operators of principal type. From Eq. (2.24) we know that $\Sigma \left(\dfrac{dx_j}{ds} \right)^2 \neq 0$ at any point on the bicharacteristics strip of any operator of principal type. Therefore, $\Sigma \left(\dfrac{dx_j}{ds} \right)^2 \neq 0$ also hold on the bicharacteristics as its projection on x space. Hence the bicharacteristics could not degenerate to a point.

The partial differential equation $Pu = f$ introduced by an operator of principal type is called **equation of principal type**. The following is the main theorem on singularity propagation of solutions to equations of principal type.

Theorem 2.8. *Assume that P is a linear partial differential operator of principal type in strict sense defined in a domain Ω with C^∞ coefficients, u is a real solution of $Pu = f$ with $f \in C^\infty(\Omega)$, $A_0 = (x_0,\xi_0) \in T^*(\Omega)$ satisfies $p_m(x_0,\xi_0) = 0$. Assume that γ is a bicharacteristic strip of P passing through A_0, then $A_0 \notin WF(u)$ implies $\gamma \cap WF(u) = \emptyset$.*

Proof. We only need to prove the conclusion locally. Because it is easy to obtain a global conclusion by gluing all local results. The proof mainly consists of two steps. The first step is to decompose the operator P, so that to reduce the problem to the case for an operator of first order, while the second step is to prove the theorem for the reduced operator of first order.

Since $\nabla_\xi p_m(x,\xi) \neq 0$, we may assume $\partial p_m/\partial \xi_n \neq 0$. Hence we may assume that $p_m(x,\xi)$ takes the form

$$p_m(x,\xi) = \xi_n^m + \sum_{k=0}^{m-1} a_{m-k}(x,\xi')\xi_n^k. \tag{2.26}$$

Since $p_m(x,\xi)$ as a polynomial of ξ_n only has simple roots at $A_0(x_0,\xi_0)$, then $p_m(x,\xi)$ can be factorized to

$$p_m(x,\xi) = (\xi_n - \lambda(x,\xi'))q_{m-1}(x,\xi), \tag{2.27}$$

where $\lambda(x,\xi')$ is a homogeneous function of ξ' with degree 1, $q_{m-1}(x,\xi)$ is a homogeneous function of ξ with degree $m-1$, and $q_{m-1}(x_0,\xi_0) \neq 0$. Next we will prove that P satisfies the factorization of operators as

$$P = (D_{x_n} - \sigma(x, D_{x'}))Q + R, \tag{2.28}$$

where R is a pseudodifferential operator, σ and Q are operators of order 1 and order $m-1$ respectively. Furthermore, the symbol of $\sigma(x, D_{x'})$ is $\lambda(x,\xi')$, the asymptotic expansion of the symbol of R is zero, while Q is a polynomial of D_{x_n}.

In order to prove Eq. (2.28) we write the symbols of σ and Q as

$$q(x,\xi',\xi_n) \sim q_{m-1} + q_{m-2} + \cdots,$$

$$\sigma(x,\xi') \sim \lambda(x,\xi') + \sigma_0 + \sigma_{-1} + \cdots,$$

where q_j is a homogeneous function of ξ with degree j, and is a polynomial of ξ_n, σ is a homogeneous function of ξ' with degree 1. To determine these symbols we use the asymptotic expansion of corresponding symbols in Eq. (2.28). Comparing the homogeneous terms with degree $m-1$ we have

$$-\sigma_0 q_{m-1} + D_{x_n} q_{m-1} - \sum_{j=1}^{n-1}(\partial_{\xi_j}\lambda)D_{x_j}q_{m-1} + (\xi_n - \lambda)q_{m-1} = p_{m-1}. \tag{2.29}$$

Let $\xi_n = \lambda(x,\xi')$, in view of $q_{m-1} \neq 0$ we have

$$\sigma_0 = \frac{1}{q_{m-1}}\left[D_{x_n}q_{m-1} - \sum_{j=1}^{n-1}(\partial_{\xi_j}\lambda)D_{x_j}q_{m-1} - p_{m-1}\right]_{\xi_n=\lambda(x,\xi')}. \tag{2.30}$$

Substituting it into Eq. (2.29) and dividing out $\xi_n - \lambda$, we can determine q_{m-2}. Successively doing in this way we can obtain the asymptotic expansion of $\sigma(x,\xi')$ and $q(x,\xi)$. In accordance, the symbols, which is the sum of these asymptotic expansions, and the corresponding pseudodifferential operators σ and Q are also obtained.

By using the factorization Eq. (2.28) the singularity propagation of the solution u to $Pu \in C^\infty$ can be reduced to the singularity propagation of the solution v satisfying $(D_{x_n} - \sigma(x, D_{x'}))v \in C^\infty$. Indeed, the facts $WF(Qu) \subset WF(u)$ and $A_0 \notin WF(u)$ imply $A_0 \notin WF(Qu)$. On the other hand, Eq. (2.28) indicates

$$(D_{x_n} - \sigma(x, D_{x'}))Qu = Pu - Ru \in C^\infty. \tag{2.31}$$

Since $q_{m-1}(x, \xi) \neq 0$, then the vector field H_{p_m} is proportional to $H_{\xi_n - \lambda}$ as $\xi_n - \lambda(x, \xi') = 0$. It means that the bicharacteristics for $p_m(x, \xi)$ and $\xi_n - \lambda(x, \xi')$ is the same. Therefore, if one can prove the theorem holds for any first order operator, i.e. $WF(Qu)$ does not meet the bicharacteristics γ of the operator $D_{x_n} - \sigma(x, D_{x'})$, then the conclusion $WF(u) \cap \gamma = \emptyset$ also holds.

Hence the problem has been reduced to the case for the solutions to first order operators. The main idea in the next step is to construct a pseudodifferential operator B of order zero with principal symbol $b_0(x, \xi)$, which does not vanish along γ, such that $Bu \in C^\infty$. Since

$$PBu = BPu + [P, B]u, \tag{2.32}$$

and $Pu \in C^\infty$. Hence if we can choose the symbol $b(x, \xi) = \sum_{j=0}^{\infty} b_{-j}(x, \xi)$ of B, such that the symbol of the commutator $[P, B]$ is zero, then the facts $[P, B]u \in C^\infty$ and $Pu \in C^\infty$ will lead to $PBu \in C^\infty$ immediately.

The requirement letting the symbol of $[P, B]$ vanish leads to a set of equalities

$$\frac{1}{i} H_{p_1} b_0 = 0,$$

$$\frac{1}{i} H_{p_1} b_{-1} + \frac{1}{i} H_{p_0} b_0 + \sum_{|\alpha|=2} \frac{1}{\alpha!} (\partial_\xi^\alpha p_1 \cdot D_x^\alpha b_0 - \partial_\xi^\alpha b_0 \cdot D_x^\alpha p_1) = 0,$$

$$\cdots\cdots$$

Since the direction of H_{p_i} is transversal to the plane $x_n = const.$, then all above equations for b_{-j} $(j \geq 0)$ are solvable, provided the initial data are given on a plane $x_n = x_{n0}$. It turns out that the problem is reduced to determine the initial data, so that $Bu \in C^\infty$ as x is sufficiently near to x_0. Indeed, if this is true, then the property of hyperbolic equations implies Bu

is C^∞. Noticing that the principal symbol of B does not vanish on γ we have $\gamma \cap WF(u) = \emptyset$, which is what we need.

The remains is to choose the initial data of b_{-j}, such that the initial data of Bu is C^∞ near $x = x_0$. We notice that the condition $(x_0, \xi_0) \notin WF(u)$ means that there is a neighborhood ω of (x_0, ξ_0) such that $\omega \cap WF(u) = \emptyset$. Choose a pseudodifferential operator A of order zero, such that the symbol $a(x, \xi)$ of A is supported in ω and is equal to 1 in a smaller neighborhood $\omega_1 \subset\subset \omega$. Then Theorem 2.6 indicates $Au \in C^\infty$. Now the initial datum of b_0 is chosen as a homogeneous function of degree 1, supported in ω_1 and is equal to 1 in a smaller neighborhood $\omega_2 \subset\subset \omega_1$. Furthermore, the initial data for b_{-j} $(j \geq 1)$ are chosen as zero. Obviously, such a choice let the symbol of $B(I - A)$ be zero as x is near to x_0. Therefore,

$$Bu = B(I - A)u + BAu \in C^\infty.$$

The proof of the theorem is thus complete. □

The theorem indicates that if A_0 is a regular point of the solution u, then all points on the bicharacteristic strip through A_0 are regular. In other words, the microlocal regularity (or singularity) of solutions to any equation of principal type propagates along bicharacteristic strip.

One may also consider the propagation of singularities in the sense of H^s. In this case the C^∞ regularity theorem of Cauchy problem for hyperbolic operator of first order used in the proof should be replaced by corresponding H^s regularity theorem. Since the energy inequality in H^s space holds for hyperbolic equation, then like Theorem 2.8 we may use similar method to prove the following theorem.

Theorem 2.9. *Assume that P is a linear partial differential operator of principal type in strict sense defined in a domain Ω with C^∞ coefficients, u is a real solution of $Pu = f$ with $f \in H^{s+m-1}(\Omega)$, $A_0 = (x_0, \xi_0) \in T^*(\Omega)$ satisfies $p_m(x_0, \xi_0) = 0$. Assume that γ is a bicharacteristic strip of P through A_0, then $A_0 \notin WF_s(u)$ implies $\gamma \cap WF_s(u) = \emptyset$.*

Remark 2.2. Denote by λ the direction $(0, \cdots, 0, \xi_1, \cdots, \xi_n)$ of the operator $\sum_j \xi_j \dfrac{\partial}{\partial \xi_j}$. If H_{p_m} is not parallel to λ, then P is called **operator of principal type in generalized sense**. The theorem 2.8 on singularity propagation can be extended to the case of the operators of principal type in generalized sense. Since the theory of Fourier integral operators has to be used to prove the corresponding theorem, we refer readers to references [74].

For Cauchy problems of partial differential equations one can also prove the conclusion on propagation of singularities of the initial data. Since the principal operators can be factorized by using the method given in Theorem 2.8, we first discuss the Cauchy problem for partial differential equations of first order:

$$\begin{cases} (D_{x_n} - \sigma(x, D_{x'}))u = f, \\ u(x', 0) = u_0(x), \end{cases} \tag{2.33}$$

where $x = (x', x_n)$, σ is a pseudodifferential operator depending on the parameter x_n with real principal symbol $\lambda(x, \xi)$.

Theorem 2.10. *Assume that $(x'_0, \xi'_0) \notin WF(u_0)$, $f \in C^\infty([0, \epsilon) \times R^{n-1}_{x'}$. Denote by γ the null characteristic strip of the symbol $\xi_n - \lambda(x, \xi')$ through $(x'_0, 0, \xi'_0, \lambda(x'_0, 0, \xi'_0))$, then*

$$\gamma \cap WF(u) = \emptyset, \quad \text{if} \quad 0 < x_n < \epsilon, \tag{2.34}$$

$$\gamma(s) \cap WF(u|_{x_n = s}) = \emptyset, \quad \text{if} \quad 0 \leq s < \epsilon, \tag{2.35}$$

Proof. We notice that the wave front set appearing in Eq. (2.35) is a subset of $R^{n-1}_{x'} \times R^{n-1}_{\xi'}$ depending on a parameter x_n. It obviously have different meaning from the wave front set in Eq. (2.34).

The proof of Theorem 2.10 is quite similar to the second part of the proof of Theorem 2.8, so we only mention the main point of it. Make an operator $B(x', D_{x'})$ such that its symbol is supported in a neighborhood of γ and does not vanish on γ. Moreover, we may let the symbol of $[D_{x_n} - \sigma(x, D_{x'}), B]$ is zero and $B(x', 0, D_{x'})u_0 \in C^\infty$. Acting B on the both sides of the equation in (2.33) we obtain

$$(D_{x_n} - \sigma(x, D_{x'}))Bu = f + [D_{x_n} - \sigma(x, D_{x'}), B]u.$$

Then the conclusion on initial boundary value problems of hyperbolic equation gives $Bu \in C^\infty$.

For any $s \in [0, \epsilon)$, the symbol of the pseudodifferential operator $B(x', s, D_{x'})$ does not vanish on $\gamma(s)$. Then from the property of wave front set we know $\gamma(s) \notin WF(u|_{x_n=s})$ as shown in Eq. (2.35).

In order to prove Eq. (2.34) we make the parametrix $A(x', x_n, D_{x'})$ of B, which also C^∞ smoothly depends on x_n. Then Theorem 2.7 implies

$$\gamma \cap WF(ABu) = \emptyset. \tag{2.36}$$

Since $ABu = u + Ru$, where Ru is a C^∞ function of x' and x_n, then u also satisfies Eq. (2.34). □

Combining with the factorization of operators we can extend the conclusion mentioned in Theorem 2.10 to the case for equations of higher order. For hyperbolic equations we have

Theorem 2.11. *Assume that P is a hyperbolic differential operator of m-th order, its principal symbol $p_m(x,\xi)$ as a polynomial of ξ_n has m simple roots $\lambda_1(x,\xi'),\cdots,\lambda_m(x,\xi')$. Assume also that u is the solution of the Cauchy problem*

$$\begin{cases} Pu = f, \\ \partial^{k-1}u|_{x_n=0} = u_{k0}(x') \quad (k=1,\cdots,m), \end{cases} \tag{2.37}$$

and $(x_0',\xi_0') \in WF(u_0), f \in C^\infty([0,\epsilon)\times R_{x'}^{n-1})$. If γ_i is the null bicharacteristic strip corresponding to the real root λ_i issuing from $(x_0',0,\xi_0',\lambda_{i0})$, where $\lambda_{i0} = \lambda(x_0',0,\xi_0')$, then for $i=1,\cdots,n$ Eq. (2.34) and Eq. (2.35) still hold with γ is replaced by γ_i.

As an application of the above theorem we can answer the problem raised in Chapter 1. For the Cauchy problem Eq. (1.35), if $\psi(x,y)$ is taken as $\delta(x,y)$, then $WF(\psi) = (0,0;\xi_0,\eta_0)$, where (ξ_0,η_0) is an arbitrary direction. Assume that (ξ_0,η_0) has been normalized, i.e. $\xi_0^2+\eta_0^2=1$, then the bicharacteristics strip through $(t,x,y,\tau,\xi,\eta) = (0,0,0,1,\xi_0,\eta_0)$ is

$$x=\xi t,\ y=\eta t,\ \tau=1,\ \xi=\xi_0,\ \eta=\eta_0, \tag{2.38}$$

whose projection on the space (t,x,y) is

$$x=\xi t,\ y=\eta t \quad \text{with } \xi^2+\eta^2=1. \tag{2.39}$$

Therefore, the singularity of the initial data at the origin propagates along all characteristic rays to the domain $t>0$. Hence the singular set of solution u is the characteristic cone with its vertex at the origin.

On the other hand, when $\psi(x,y)=\theta(x)$, $WF(\psi) = (0,y,\xi,0)$, where ξ can be any real number (in fact it is enough to be ± 1 due to homogeneity). According to Theorem 2.11 the projections of all bicharacteristic strip starting from the points in $WF(\psi)$ are $\{x=\pm t,\ -\infty<y<\infty\}$. They form the plane $x=\pm t$. These bicharacteristics do not enter inside the wedge $-t<x<t$, then $u_\theta(t,x,y)$ does not have singularity inside the wedge. It turns out that the conclusion we obtained by using the general theory of singularity propagation proved in this Chapter coincides with the phenomena observed in Chapter 1.

2.3 Reflection of singularity on boundary

In the last section we discussed the propagation of singularity of a solution to partial differential equations inside a domain, where the solution is defined. In the study of the solutions in a domain with boundary, the bicharacteristics may meet the boundary. In this case how does the distribution of singularities behave near the boundary? This is the problem of **reflection of singularity**. In this case the distribution of singularities near the boundary is obviously related to the setting of the boundary conditions, and is also related to the behavior of the intersection of the bicharacteristics with the boundary. In this section we mainly consider the case when the bicharacteristics transversally intersects with boundaries.

Suppose that the boundary is C^∞ smooth, then one can introduce a C^∞ homomorphism near the boundary to flatten the boundary. Since the type of the partial differential equations are invariant under any homomorphism. The characteristics, the bicharacteristics and the wave front set are transformed obeying the rule induced by the homomorphism, then we may only consider the case when the boundary has been flatten without loss of generality. In the sequel we always denote the domain, where the solution is defined, by $x_n > 0$, and denote the boundary by $x_n = 0$.

Next we consider the reflection of singularity for the system of partial differential equations of first order, because any partial differential equation of higher order can be easily reduced to the case of the system of equations of first order.

Consider the system

$$(D_{x_n} - A(x, D'_x))U = F, \tag{2.40}$$

where $x' = (x_1, \cdots, x_n)$, A is an $N \times N$ matrix of pseudodifferantial operators of first order, C^∞ smoothly depending on x_n. The principal symbol $a(x, \xi)$ of A has real eigenvalues $\lambda_1, \cdots, \lambda_n$ different from each other. U and F are vectors with N components, and F are C^∞. On the boundary $x_n = 0$, U satisfies

$$B(x', D'_x)U = h, \tag{2.41}$$

where B is a $k \times N$ matrix of pseudodifferantial operators of order zero, h is a C^∞ vector function with k components. Next we can see that the number k is closely related to the sign of eigenvalues $\lambda_1, \cdots, \lambda_n$.

In order to discuss the singularity of the solution to (2.40), (2.41) near $x_n = 0$ we first reduce the system Eq. (2.40) to a diagonal-like form, then the result on Cauchy problem of single equation can be employed. To this end we prove some lemmas.

Lemma 2.1. *Assume that E, F are $N_1 \times N_1$,$N_2 \times N_2$ matrices respectively, the eigenvalues of E and F are separated from each other, then $\phi(T) = TF - ET$ is an injective and surjective map.*

Proof. In fact we only need to prove that $\phi(T) = TF - ET$ is injective, i.e. $TF - ET = 0$ implies $T = 0$.

Suppose that E has decomposed to the form $\mathrm{diag}(E_1, \cdots, E_n)$, where each E_j is a Jordan block of order ν_j:

$$E_j = \begin{pmatrix} \lambda_j & & & \\ 1 & \lambda_j & & \\ & \ddots & \ddots & \\ & & 1 & \lambda_j \end{pmatrix}.$$

Denote the rows of T by $T_1, \cdots, T_n$,

$$TF = ET = \begin{pmatrix} \lambda_1 T_1 \\ T_1 + \lambda_1 T_2 \\ \cdots\cdots \\ T_{\nu_1 - 1} + \lambda_1 T_{\nu_1} \\ \cdots\cdots \end{pmatrix}.$$

Hence we have $T_1 F = \lambda_1 T_1$. Since λ_1 is not eigenvalue of F, then $T_1 = 0$. Next we can use similar method to know $T_2 = \cdots = T_{\nu_1} = 0$. Continue in this fashion we know every row of T is zero.

For general matrix E, we find a matrix L such that $\tilde{E} = LEL^{-1}$ has a Jordan standard form. Since $TF = (L^{-1}\tilde{E}L)T$, then $(LT)F = \tilde{E}(LT)$. According to the previous proof we know $LT = 0$. This leads to $T = 0$. □

Lemma 2.2. *For the system of pseudodifferential equations*

$$D_{x_n} V = GV + HV, \tag{2.42}$$

where $G = \mathrm{diag}(E, F)$ has a homogeneous symbol of degree 1 with respect to ξ', E and F are $N_1 \times N_1$ and $N_2 \times N_2$ matrices of pseudodifferential operators, whose symbols have eigenvalues different from each other, H is a pseudodifferential operator of order 0, then there is a transformation $W = SV$, such that W satisfies

$$D_{x_n} W = GW + \alpha W + RW, \tag{2.43}$$

where $\alpha = \mathrm{diag}(\alpha_1, \alpha_2)$ is a matrix of pseudodifferential operators of order 0, R is a matrix pseudodifferential operators of order $-\infty$.

Proof. Let $W^{(1)} = (I+K_1)V$, where K_1 is a matrix of pseudodifferential operators of order -1, then

$$\begin{aligned} D_{x_n}W^{(1)} &= (I+K_1)GV + (I+K_1)HV + D_{x_n}K_1V \\ &= (I+K_1)(G+H)(I+K_1)^{-1}W^{(1)} + \cdots \\ &= GW^{(1)} + (K_1G - GK_1 + H)W^{(1)} + \cdots, \end{aligned}$$

where "$\cdots$" means the term obtained by acting a pseudodifferential operator of order -1 on $W^{(1)}$.

Next we are going to look for matrices A_1, A_2 and K_1, such that

$$K_1G - GK_1 + H = \operatorname{diag}(A_1, A_2) \tag{2.44}$$

holds. Denote $H = \begin{pmatrix} H_{11} & H_{12} \\ H_{21} & H_{22} \end{pmatrix}$, and take K_1 as a matrix with the form $\begin{pmatrix} 0 & K_{12} \\ K_{21} & 0 \end{pmatrix}$, then choose $A_1 = H_{11}, A_2 = H_{22}$. Since the eigenvalues of E and F are different from each other, then according to Lemma 2.1 we can find matrices K_{12}, K_{21}, such that

$$K_{12}F - EK_{12} = -H_{12}, \quad K_{21}E - FK_{21} = -H_{21}, \tag{2.45}$$

then $W^{(1)}$ satisfies

$$D_{x_n}W^{(1)} = GW^{(1)} + \operatorname{diag}(A_1, A_2)W^{(1)} + BW^{(1)}, \tag{2.46}$$

where B is a matrix of pseudodifferential operators of order -1.

Let $W^{(2)} = (I+K_2)W^{(1)}$, where K_2 is a matrix of pseudodifferential operators of order -2, then

$$D_{x_n}W^{(2)} = GW^{(2)} + \operatorname{diag}(A_1, A_2)W^{(2)} + (K_2G - GK_2 + B)W^{(2)} + \cdots,$$

where "$\cdots$" stands for a term obtained by acting a pseudodifferential operators of order -2 on $W^{(2)}$. Like the above calculations, the matrices B and K_2 can be determined, such that

$$D_{x_n}W^{(2)} = GW^{(2)} + \operatorname{diag}(A_1, A_2)W^{(2)} + \operatorname{diag}(B_1, B_2)W^{(2)} + \cdots,$$

Going on in this way, the order of the remaining pseudodifferential operators will be lower and lower. Then similar to the process of constructing a pseudodifferential operator by using its asymptotic expansion of symbol we can construct $I + K$ by using an infinite product

$$I + K \sim \prod_{\ell=1}^{\infty}(I + K_\ell).$$

Set $W = (I+K)V$, we obtain Eq. (2.43). □

Lemma 2.3. *For the system of pseudodifferential equations*

$$(D_{x_n} - A(x, D_{x'}))U = F, \tag{2.47}$$

where A is an $N \times N$ matrix of pseudodifferential operators C^∞ smoothly depending on x_n, whose principal symbol $a9x, \xi')$ has real eigenvalues $\lambda_1(x,\xi'), \cdots, \lambda_N(x,\xi')$, different from each other, then there is a pseudodifferential operator $S(x, D_x)$, such that $W = SU$ satisfies

$$D_{x_n} W = \mathrm{diag}(\sigma_1, \cdots, \sigma_N)W + RW + SF, \tag{2.48}$$

where $\sigma_j = \sigma_j(x, D'_x)$ has principal symbol $\lambda_j(x,\xi')$, R is a matrix of pseudodifferential operator of $-\infty$.

Proof. According to the property of $a(x,\xi')$ we can find a matrix $e(x,\xi')$, homogeneous of degree 0 with respet to ξ', such that

$$e \cdot a \cdot e^{-1} = \tilde{a} = \mathrm{diag}(\lambda_1, \cdots, \lambda_N).$$

Then, let $V = EU = e(x, D_{x'})U$, we have

$$D_{x_n} V = \tilde{A}(x, D_{x'})V + HV + EF, \tag{2.49}$$

where H is a pseudodifferential operator of order 0. Applying Lemma 2.2 we can reduce Eq. (2.49) to the form Eq. (2.48) by virtue of $W = S_1 V$. Then $S = S_1 E$ leads to the required conclusion. □

Combining the above lemmas we can prove the following theorem on regularity reflection of solutions to system of hyperbolic equations.

Theorem 2.12. *For the problem (2.40), (2.41), assume that $\lambda_1, \cdots, \lambda_n$ are N real eigenvalues of the symbol $a(x,\xi')$ of the operator A, different from each other. Denote by $\gamma_j = (x_j(s), \xi_j(s))$ the null bicharacteristic strip of the symbol $\xi_n - \lambda_j(x,\xi')$ through $(x'_0, 0; \xi'_0, \lambda_j(x'_0, 0, \xi'_0))$, S is the matrix of operators introduced in Lemma 2.3, which transform the system (2.40) into an almost diagonal form, which is different from a diagonal matrix by an operator matrix of order $-\infty$. Assume that on $x_n = x_{n0}$ the equality $WF(U) \cap \gamma_j = \emptyset$ holds for $j = j_1, \cdots, j_{N_1}$. Moreover, assume that BS^{-1} is elliptic with respect to $W_{i_1}, \cdots, W_{i_{N-N_1}}$ on $x_n = 0$, where*

$$\{i_1, \cdots, i_{N-N_1}\} = \{1, \cdots, N\} \setminus \{j_1, \cdots, j_{N_1}\},$$

then $WF(U) \cap \gamma_j = \emptyset$ holds for any j.

Proof. Since $W = SU$ is an invertible transformation, then the conclusion in the theorem is equivalent to that $WF(U) \cap \gamma_j = \emptyset$ holds for any j. By using Lemma 2.3 we can diagonalize the system (2.40), so that W satisfies

$$D_{x_n} W = \text{diag}(\sigma_1, \cdots, \sigma_N)W + C^\infty \text{ terms.} \tag{2.50}$$

Notice that the C^∞ terms do not influence the C^∞-singularity analysis, then we can treat all components of W separately.

Since the operator $D_{x_n} - \sigma_i(x, D_{x'})$ is elliptic on γ_j as $i \neq j$, then Eq. (2.50) means $WF(W_i) \cap \gamma_j = \emptyset$ as $i \neq j$. Now by taking the plane $x_n = x_{n0}$ as the initial plane, we know from the classical result on singularity propagation for Cauchy problem of hyperbolic equation and the assumptions of the theorem

$$(x_0', \xi_0') \notin \bigcup_{s=1}^{N_1} WF(W_{j_s}|_{x_n=0}). \tag{2.51}$$

In view of the ellipticity of BS^{-1} with respect to $W_{i_1}, \cdots, W_{i_{N-N_1}}$, then from $BS^{-1}W = h \in C^\infty$ we obtain

$$(x_0', \xi_0') \notin \bigcup_{s=1}^{N-N_1} WF(W_{j_s}|_{x_n=0}). \tag{2.52}$$

Now use the the value of W on $x_n = 0$ as initial value, and again use Theorem 2.3 we can obtain $WF(W_{i_s}) \cap \gamma_{i_s} = \emptyset$ holds for all $1 \leq s \leq N - N_1$. It means that $WF(W)$ does not meet any γ_s, hence $WF(U)$ also does not meet any γ_s. □

Remark 2.3. The assumptions in Theorem 2.5 contains the requirement of the transversal intersection of bicharacteristics with the boundary. Indeed, γ_i is the integral curve of the Hamilton vector field of $\xi_n - \lambda_i(x, \xi')$. Hence

$$\frac{dx_n}{dt} = \frac{\partial}{\partial \xi_n}(\xi_n - \lambda_i(x, \xi')) \neq 0,$$

then the projection of γ_i on the base space is transversal to the boundary $x_n = 0$.

In the above theorem the operator in Eq. (2.40) is required to have N real eigenvalues. It means that the system is hyperbolic with respect to x_n, so that the problem (2.40), (2.41) is an initial problem in fact. The requirement of having N real eigenvalues could not be satisfied even for general boundary value problem of a hyperbolic system. For instance, for the wave equation defined in $(0,T) \times \Omega$, this condition on the boundary $(0,T) \times \partial\Omega$ is not satisfied either. Next we will give a more general theorem on singularity reflection without this requirement.

Consider the boundary value problem which still takes the form Eq. (2.40), (2.41) but the matrix $a(x,\xi')$ is assumed to have N_0 simple real eigenvalues, m^+ complex eigenvalues with positive imaginary part, m^- complex eigenvalues with negative imaginary part in a domain ω of (x_0, ξ_0'). Like the discussion given above we first have to block-diagonalize the system. Since Lemma 2.1 is also valid for the matrix with complex eigenvalues, then we have the following lemma

Lemma 2.4. *If U is a solution of Eq. (2.40), the eigenvalues of the symbol $a(x,\xi')$ of the matrix of pseudodifferential operators $A(x, D_{x'})$ are given as above, then there is a matrix $S(x, D_{x'})$, such that $W = SU$ satisfies*

$$D_{x_n} W = HW + RW, \tag{2.53}$$

where

$$H = \begin{pmatrix} \sigma_1(x, D_{x'}) & & & & \\ & \ddots & & & \\ & & \sigma_{N_0}(x, D_{x'}) & & \\ & & & e^+(x, D_{x'}) & \\ & & & & e^-(x, D_{x'}) \end{pmatrix}.$$

The principal symbol of $\sigma_j(x, D_{x'})$ is $\lambda_j(x,\xi')$, while $e^+(x, D_{x'}), e^-(x, D_{x'})$ have complex symbols of first order, R is a matrix of pseudodifferential operators of order $-\infty$.

Therefore, in order to derive a result on singularity reflection for general cases we have to discuss the regularity of solutions to boundary value problems for elliptic systems.

Theorem 2.13. *Assume that W^+ satisfies*

$$\begin{cases} D_{x_n} W^+ = E^+ W^+ + F^+; \\ W^+(0) = h^+, \end{cases} \tag{2.54}$$

where E^+ has its principal symbol $e^+(x,\xi')$ satisfying

$$Im(\mathrm{spec}(e^+)) \geq c_0|\xi|, \quad c_0 > 0. \tag{2.55}$$

Assume more $(x_0', \xi_0') \notin WF(h^+)$ and $(x_0', \xi_0') \notin WF(F^+|_{x_n=s})$ for $0 \leq s \leq \epsilon$, then $(x_0', \xi_0') \notin WF(W^+|_{x_n=s})$ for $0 \leq s \leq \epsilon$ is also valid.

Proof. Let us first consider the case $F^+ = 0$. We are going to prove that there is a pseudodifferential operator $B = b(x, D_{x'})$ of order zero, so that W^+ can be written as the form $b(x, D_{x'})h^+$, i.e. $W^+ = Bh^+$.

Assume that the symbol of the operator $b(x, D_{x'})$ has an asymptotic expansion $b \sim \sum_{j=0}^{-\infty} b_j$, such that each term b_j is a homogeneous symbol of order j, and $\partial_{x_n}^k b_j$ is a homogeneous symbol of order $j + k$. To determine b_j, we substitute $W^+ = Bh^+$ into the equation

$$D_{x_n} W^+ = e^+(x, D_{x'})W^+.$$

The rule of the calculation for pseudodifferential operators gives:

$$\begin{aligned} &D_{x_n} b_0 - e_1^+ b_0 = 0, \\ &D_{x_n} b_{-1} - e_1^+ b_{-1} = \sum_{|\alpha|=1} \partial_\xi^{(\alpha)} e_1^+ \cdot D_x^\alpha b_0 + e_0^+ b_0, \\ &\cdots\cdots \end{aligned} \tag{2.56}$$

Taking the initial data for these symbols as $b_0 = 1, b_{-j} = 0 \ \ (j > 0)$, then all b_j with $j \geq 0$ can be determined. For instance,

$$b_0 = \exp\left(i \int_0^{x_n} e_1^+ dx_n \right).$$

It is easy to check that $\partial_{x_n}^k b_0$ is a symbol of order k. Similarly, $\partial_{x_n}^k b_j$ is a symbol of order $k + j$. This means that W^+ can be written as $b(x, D_{x'})h^+$.

Now let us consider the case $F^+ \neq 0$. In this case W^+ can be written as

$$W^+ = b(x, D_{x'})h^+ + \int_0^{x_n} \tilde{b}(x', s, D_{x'})h^+ F^(x', s)ds, \tag{2.57}$$

where $\tilde{b}(x', s, D_{x'})$ is the solution operator of the Cauchy problem

$$\begin{cases} D_{x_n} W^+ = e^+(x, D_{x'})W^+, \\ W^+|_{x_n = s} = F^+(x', s). \end{cases} \tag{2.58}$$

It is also a pseudodifferential operator of order zero, C^∞ depending on x_n and s. Hence the conclusion of the theorem also holds for $F' \neq 0$. □

Remark 2.4. Inverse the direction x_n in Theorem 2.13, one can obtain a similar conclusion for another elliptic operator. For the problem

$$\begin{cases} D_{x_n} W^- = E^- W^- + F^-, \\ W^-(\epsilon) = h^-, \end{cases} \tag{2.59}$$

where E^- has its principal symbol $e^-(x,\xi')$ satisfying

$$Im(\text{spec}(e^-)) \le c_1|\xi|, \quad c_1 < 0. \tag{2.60}$$

Moreover, if $(x_0', \xi_0') \notin WF(h^-)$, and if $(x_0', \xi_0') \notin WF(F^-|_{x_n=s})$ for $0 \le s \le \epsilon$, then $(x_0', \xi_0') \notin WF(W^-|_{x_n=s})$ for $0 \le s \le \epsilon$.

Combining the above discussion on the hyperbolic case and the elliptic case we have the following more general conclusion.

Theorem 2.14. *For the problem (2.40), (2.41), assume that in a neighborhood ω of (x_0', ξ_0') the symbol $a(x,\xi')$ of the operator A has N_0 simple real eigenvalues $\lambda_1, \cdots, \lambda_{N_0}$, m^+ complex eigenvalues with positive imaginary part, m^- complex eigenvalues with negative imaginary part. Denote by $\gamma_j = (x_j(s), \xi_j(s))$ the null bicharacteristic strip of the symbol $\xi_n - \lambda_j(x,\xi')$ through $(x_0', 0; \xi_0', \lambda_j(x_0', 0, \xi_0')$, S is the matrix of operators introduced in Lemma 2.3, which transform the system (2.40) into the almost diagonal form. Assume that on $x_n = x_{n0}$ $(0 < x_{n0} < \epsilon)$ the equality $WF(U) \cap \gamma_j = \emptyset$ holds for $j = j_1, \cdots, j_{N_1}$. Moreover, assume that $(x_0', \xi_0') \notin WF(h)$, BS^{-1} is elliptic with respect to $W_{i_1}, \cdots, W_{i_{N_0-N_1}}$, W^+ is elliptic, where*

$$\{i_1, \cdots, i_{N_0-N_1}\} = \{1, \cdots, N\} \setminus \{j_1, \cdots, j_{N_1}\},$$

then $WF(U) \cap \gamma_j = \emptyset$ holds for any $j \le N_0$.

Proof. By using the transformation $W = SU$ the system (2.40) is reduced to block-diagonal form. Since

$$WF(U|_{x_n=x_{n0}}) \bigcap \left(\bigcup_{s=1}^{N_1} \gamma_{j_s} \right) = \emptyset$$

on $x = x_{n0}$, then we also have

$$WF(W|_{x_n=x_{n0}}) \bigcap \left(\bigcup_{s=1}^{N_1} \gamma_{j_s} \right) = \emptyset$$

Taking $x_n = x_{n0}$ as initial plane, Theorem 2.12 on Cauchy problems of hyperbolic system implies

$$(x_0', \xi_0') \notin \bigcup_{s=1}^{N_1} WF(W_{j_s})|_{x_n=0}),$$

Then the remark behind Theorem 2.13 gives

$$(x_0', \xi_0') \notin \bigcup_{s=1}^{N_1} WF(W^-)|_{x_n=0}).$$

Hence W_{j_s} $(s = 1, \cdots, j)$ and W^- on $x_n = 0$ are microlocal regular at (x_0, ξ_0'). According to the assumptions on the ellipticity of BS^{-1} on $x_n = 0$ of this theorem we have

$$(x_0', \xi_0') \notin WF(W^+|_{x_n=0}) \bigcup \left(\bigcup_{s=1}^{N_0-N_1} WF(W_{i_s}|_{x_n=0}) \right).$$

Again consider the Cauchy problem with $x_n = 0$ as initial plane. It is easy to see that Theorems 2.12 and 2.13 imply

$$\bigcup_{s=1}^{N_0-N_1} \gamma_{i_s} \bigcap WF(W) = \emptyset.$$

Since S is elliptic at a neighborhood of (x_0', ξ_0'), then

$$\bigcup_{s=1}^{N_0-N_1} \gamma_{i_s} \bigcap WF(U) = \emptyset.$$

The theorem is thus proved. □

Example 1 Next we take the wave equation

$$u_{tt} - c^2(u_{xx} + u_{yy} + u_{zz}) = 0 \tag{2.61}$$

in the domain $x > 0$ as example to study the singularity reflection of its solution near the boundary $x = 0$.

The symbol of the wave operator is $\tau^2 - c^2(\xi^2 + \eta^2 + \zeta^2)$. Let Λ be a pseudodifferential operator with the symbol $\lambda = (\xi^2 + \eta^2 + \zeta^2)^{1/2}$, $U = (\Lambda u, D_x u)$, then Eq. (2.61) can be reduced to a system of first order as

$$D_x U = A(D_x, D_y, D_z)U, \tag{2.62}$$

where

$$A = \begin{pmatrix} 0 & \Lambda \\ \Lambda^{-1}(c^{-2}D_t^2 - D_y^2 - D_z^2) & 0 \end{pmatrix}.$$

Its principal symbol is

$$a = \begin{pmatrix} 0 & \lambda \\ \lambda^{-1}(c^{-2}\tau^2 - \eta^2 - \zeta^2) & 0. \end{pmatrix}.$$

Obviously, the eigenvalue of a is $\pm(c^{-2}D_t^2 - D_y^2 - D_z^2)^{1/2}$. For the point $(t, y, z, \tau, \eta, \zeta)$ on $T^*(R^3_{tyz})$ we have:

If $\tau < c(\eta^2 + \zeta^2)^{1/2}$, then for any ξ, the symbol $\tau^2 - c^2(\xi^2 + \eta^2 + \zeta^2) \neq 0$, then u is microlocally regular at (τ, ξ, η, ζ). Hence $u|_{x=s}$ is microlocally regular for any $s > 0$. By using the fact that the eigenvalue of a is imaginary as $\tau < c(\eta^2 + \zeta^2)^{1/2}$, then $u|_{x=0}$ is also microlocally regular at (τ, η, ζ).

If $\tau > c(\eta^2+\zeta^2)^{1/2}$, then a has two real eigenvalues $\xi_\pm = \pm(\tau^2 - c^2(\eta^2 + \zeta^2))^{1/2}$. Correspondingly, one can draw two bicharacteristic strips γ^+, γ^-. If for sufficiently small x, $WF(u) \cap \gamma^+ = \emptyset$, then Theorem 2.14 implies $WF(u) \cap \gamma^- = \emptyset$ and vice versa. In one words, if u has singularity on one of the two bicharacteristic strips $\gamma^\pm$, then the singularity mast be reflected to another bicharacteristic strip.

Making projection of the bicharacteristic strips on the base space (t, x, y, z) we obtain the property of the reflection of ray of light on the boundary $x = 0$. Indeed, if the incident line is

$$x = \xi_+ t,\ y = \eta t + y_0,\ z = \zeta t + z_0;\ t < 0,$$

then the reflect line is

$$x = \xi_- t,\ y = \eta t + y_0,\ z = \zeta t + z_0;\ t > 0.$$

Because $\xi_+ = -\xi_-$, the vectors (ξ_+, η, ζ), (ξ_-, η, ζ) and the normal $(1,0,0)$ locate on a same plane. Meanwhile, the angle between $(\xi_\pm, \eta, \zeta)$ and $(1,0,0)$ are equal. The fact coincides the classical conclusion in the law of reflection of light: the reflective ray locates on the plane formed by the incident ray with the normal and is in the side different from the incident ray, the reflect angle is equal to the incident angle.

Example 2 Consider the reflection and refraction of light on an interface of two different media. Suppose that two different media are placed on both sides of the plane $x = 0$. The speed of propagation of light in the domains $\pm x > 0$ is $c_\pm$, then the partial differential equation describing the propagation of light is

$$\begin{cases} u_{tt} - c_+^2(u_{xx} + u_{yy} + u_{zz}) = 0, & x > 0 \\ u_{tt} - c_-^2(u_{xx} + u_{yy} + u_{zz}) = 0, & x < 0. \end{cases} \tag{2.63}$$

It is required that u and its derivatives are continuous on $x = 0$. Now if u has singularity on a bicharacteristics in the "+" medium, which arrives at the origin from the domain $x > 0$ for $t = 0$, how does the singularity propagate in future?

All bicharacteristics from the domain $x > 0$ arriving at the origin at $t = 0$ forms a half of the characteristic cone $c_+ t = -\sqrt{x^2 + y^2 + z^2}$ $(x > 0)$. Assume that the projection of the bicharacteristic strip ℓ bearing singularity on the space (x, y, z) is the straight line L, which locates on the plane $z = 0$ arrives at the origin and forms an angle θ with y-axis. The equation of L is $y = -x \tan\theta,\ z = 0$ in the space (x, y, z) or

$$x = -\frac{c_+ t}{\sqrt{1+\tan^2\theta}},\quad y = \frac{(c_- \tan\theta)t}{\sqrt{1+\tan^2\theta}},\quad z = 0 \tag{2.64}$$

in the space (t, x, y, z). In order to solve this problem we fold the whole space (t, x, y, z) to the upper half space $x > 0$. Set

$$\tilde{u}(t, x, y, z) = (\tilde{u}_1, \tilde{u}_2) = {}^t(u(t, x, y, z), u(t, -x, y, z)) \tag{2.65}$$

for $x > 0$, then $\tilde{u}(t, x, y, z)$ satisfies

$$\tilde{u}_{1xx} = c_+^{-2}\tilde{u}_{1tt} + \tilde{u}_{1yy} + \tilde{u}_{1zz};$$

$$\tilde{u}_{2xx} = c_-^{-2}\tilde{u}_{1tt} + \tilde{u}_{1yy} + \tilde{u}_{1zz}.$$

While the consistency condition on the interface is reduced to

$$\tilde{u}_1(t, 0, y, z) = \tilde{u}_2(t, 0, y, z)$$

$$D_x\tilde{u}_1(t, 0, y, z) = D_x\tilde{u}_2(t, 0, y, z).$$

Denote by Λ the pseudodifferential operator with the symbol $\lambda = (\tau^2 + \eta^2 + \zeta^2)^{1/2}$, and denote

$$U = {}^t(\Lambda\tilde{u}_1, , D_x\tilde{u}_1, \Lambda\tilde{u}_2, , D_x\tilde{u}_2),$$

then

$$D_x U = A(D_t, D_y, D_z)U, \tag{2.66}$$

where

$$A = \begin{pmatrix} 0 & \Lambda & 0 & 0 \\ E_+ & 0 & 0 & 0 \\ 0 & 0 & 0 & \Lambda \\ 0 & 0 & E_- & 0 \end{pmatrix},$$

$$E_\pm = \Lambda^{-1}(c_\pm^{-2}D_t^2 - D_y^2 - D_z^2).$$

It is easy to compute that the eigenvalue of the symbol a of A is $\pm(c_\pm^{-2}\tau^2 - \eta^2 - \zeta^2)^{1/2}$.

The boundary condition of the problem with U as its unknown function can be written as the form

$$BU = 0, \tag{2.67}$$

where

$$B = \begin{pmatrix} 1 & 0 & -1 & 0 \\ 0 & 1 & 0 & 1 \end{pmatrix}. \tag{2.68}$$

Now let us apply Theorem 2.12 and Theorem 2.14 to the problem. First, we look for the matrix S^{-1} and S, which can diagonalize the system (2.66). The symbols of these two matrices of operators are

$$\begin{pmatrix} \lambda & \lambda & 0 & 0 \\ (\lambda e_+)^{1/2} & -(\lambda e_+)^{1/2} & 0 & 0 \\ 0 & 0 & \lambda & \lambda \\ 0 & 0 & (\lambda e_+)^{1/2} & -(\lambda e_+)^{1/2} \end{pmatrix},$$

$$\begin{pmatrix} \frac{1}{2}\lambda^{-1} & \frac{1}{2}(\lambda e_+)^{-1/2} & 0 & 0 \\ \frac{1}{2}\lambda^{-1} & -\frac{1}{2}(\lambda e_+)^{-1/2} & 0 & 0 \\ 0 & 0 & \frac{1}{2}\lambda^{-1} & \frac{1}{2}(\lambda e_-)^{-1/2} \\ 0 & 0 & \frac{1}{2}\lambda^{-1} & -\frac{1}{2}(\lambda e_-)^{-1/2}, \end{pmatrix},$$

where $e_\pm = \lambda^{-1}(c_\mp^{-2}\tau^2 - \eta^2 - \zeta^2)$. Then the principal symbol of SAS^{-1} is

$$\mathrm{diag}(e_+, -e_+, e_-, -e_-). \tag{2.69}$$

On the bicharacteristic strip

$$\frac{dt}{ds} = \tau, \quad \frac{dx}{ds} = c_\pm^2 \xi, \quad \frac{dy}{ds} = c_\pm^2 \eta, \quad \frac{dz}{ds} = c_\pm^2 \zeta,$$

where τ, ξ, η, ζ are constant, satisfying $\tau^2 - c^2(\xi^2 + \eta^2 + \zeta^2) = 0$. By the homogeneity we may take $\tau = 1$ in the sequel. Hence on the bicharacteristic strip ℓ

$$\xi = -\frac{c_+^{-1}}{\sqrt{1+\tan^2\theta}}, \quad \eta = -\frac{c_+^{-1}\tan\theta}{\sqrt{1+\tan^2\theta}}, \quad \zeta = 0.$$

For $(\xi, \eta, \zeta) = (1, c_+^{-1}\tan\theta(\sqrt{1+\tan^2\theta})^{-1/2}, 0)$, the four eigenvalues of the matrix a is

$$\frac{\pm c_+^{-1}}{\sqrt{1+\tan^2\theta}}, \quad \pm\left(c_-^{-2} - c_+^{-2}\frac{\tan^2\theta}{1+\tan^2\theta}\right)^{1/2}.$$

When $c_+^2 > c_-^2 \tan^2\theta(1+\tan^2\theta)^{-1/2}$, these four eigenvalues are all real. Then the projection of bicharacteristic strips through the origin on the space (x, y, z) is

$$L_1: \ y = x\tan\theta,$$

$$L_2: \ y = -x\tan\theta,$$

$$L_3: \ y = \frac{c_+^{-1} x \tan\theta}{\sqrt{c_-^{-2} + (c_-^{-2} - c_+^{-2})\tan^2\theta}},$$

$$L_4: \ y = \frac{-c_+^{-1} x \tan\theta}{\sqrt{c_-^{-2} + (c_-^{-2} - c_+^{-2})\tan^2\theta}},$$

where L_2 is the equation of the incident ray.

According to the notations in Theorem 2.14, we take $i_1 = 1, i_2 = 3$, then

$$BS^{-1} = \left(\frac{\lambda}{\sqrt{\lambda e_+}} \; \frac{\lambda}{-\sqrt{\lambda e_+}} \; \frac{-\lambda}{\sqrt{\lambda e_-}} \; \frac{-\lambda}{-\sqrt{\lambda e_-}} \right).$$

The first column and the third column of the matrix forms a submatrix with rank 2. Since L_1, L_3 are rays issuing from the origin to $y > 0$, the solution of Eq. (2.66) is regular on L_1 and L_3. Then according to the conclusion of Theorem 2.12 the solution must be regular on L_2. Conversely, if U has singularity on L_2, it must propagate to $L_1$and L_3. By using the inverse of the folding transformation introduced above and going back to the original equation (2.63), we see that L_1 is the reflective ray of L_2, and the image

$$L_3' : y = \frac{-c_+^{-1} x \tan\theta}{\sqrt{c_-^{-2} + (c_-^{-2} - c_+^{-2}) \tan^2\theta}}$$

of L_3 is the refractive ray.

Denote by ψ the angle between L_3' and the x-axis, then

$$\frac{c_+^{-1} \tan\theta}{\sqrt{c_-^{-2} + (c_-^{-2} - c_+^{-2}) \tan^2\theta}} = \tan\psi,$$

which implies

$$c_-^{-2} \cot^2\theta + c_-^{-2} - c_+^{-2} = c_+^{-2} \cot^2\psi.$$

Hence we have

$$\frac{\sin\theta}{\sin\psi} = \frac{c_+}{c_-}. \tag{2.70}$$

This coincides the law of refraction of ray of light.

Remark 2.5. When $c_-^{-2} < c_+^{-2} \tan^2\theta/(1 + \tan^2\theta)$, there are two complex eigenvalues among the four eigenvalues of the matrix a. In this case the singularity on the incident ray only propagates along the reflective ray and no refractive ray appears. This corresponds to the phenomena of total reflection in geometric optics.

2.4 Further discussions

The problems discussed in the above sections are fundamental problems in the theory of singularity analysis for linear partial differential equations. The development of the theory raised many significant and important problems, which are usually more difficult, so that more careful calculations are required. For instance, the following problems attracted many mathematicians attention.

2.4.1 *Generalized reflection of singularity on boundary*

In the discussion of the singularity reflection in Section 2.3 we always assume that the bicharacteristic ray bearing the singularities of solutions intersects the boundary transversally. However, it is also possible that the bicharacteristic ray is tangential to the boundary, when the ray meets the boundary. For instance, if ω is a convex domain in R^n, and consider the propagation of singularity of solution to wave equation in the domain $(R^n \backslash \omega) \times (-\infty, \infty)$, then the bicharacteristics of solutions can be tangential to the boundary. Such a case is called **glancing**. When glancing occurs, the discussion in the above section does not work. For instance, the symbol of the operator A of the system (2.40) does not have n bounded eigenvalues. Therefore, the singularity analysis near the boundary in the glancing case will be more complicated.

Assume that M is a manifold with boundary, consider the behavior of the singularity of the solution to

$$\begin{cases} Pu \in C^\infty(M), \\ Bu \in C^\infty(\partial M). \end{cases} \tag{2.71}$$

Assume that P is an operator of principal type of second order, the boundary ∂M is non-characteristic, Bu in the boundary condition takes the form u or $\dfrac{\partial u}{\partial \nu} + \beta u$, where ν is the conormal direction. In order to describe the behavior of the singularity of solution, we introduce some notations. Next we assume that a simplified coordinate system has been introduced in the manifold, so that the boundary ∂M has been flattened to $x = 0$, so that M is $x \geq 0$.

Introduce an equivalence relation " $\sim$ " in T^*M: in the case when two points $z_1, z_2 \in T^*M$, and at least one of them is in the set of inner points of T^*M, we say $z_1 \sim z_2$ if and only if $z_1 = z_2$; while in the case

$z_1, z_2 \in \partial T^*M \setminus N^*\partial M$ we say $z_1 \sim z_2$ if and only if the projection on $T^*(\partial M) \setminus 0$ is the same.

Let $DM = ((T^*M \setminus 0) \setminus N^*(\partial M))/\sim$, then one can introduce a projection $b : T^*M \to DM$. Denote the image of the characteristic set $\Sigma(P) = p_2^{-1}(0)$ of the operator P of the projection on DM by Σ_b can be written as

$$\Sigma_b^0 \cup \Sigma_b^1 \cup \Sigma_b^2 \cup \cdots \cup \Sigma_b^\infty, \tag{2.72}$$

where

$$\Sigma_b^0 = \Sigma_b \bigcap T^*(M) \setminus 0;$$
$$\Sigma_b^1 = \{\rho \in \Sigma_b,\ b^{-1} \text{ contains two points}\};$$
$$\Sigma_b^k = \{\rho \in \Sigma_b \setminus (\Sigma_b^0 \cup \Sigma_b^1); H_{p_2}^j x = 0 \text{ for all } j \le k-1,\ H_{p_2}^k \neq 0\};$$
$$\Sigma_b^\infty = \{\rho \in \Sigma_b \setminus (\Sigma_b^0 \cup \Sigma_b^1); H_{p_2}^j x = 0 \text{ for all } j\}.$$

Denote $\Sigma_b^{(k)} = \bigcup_{j \ge k} \Sigma_b^j$, then $G = \Sigma_b^{(2)}$ is the set of glancing points. On Σ_b^2 the point with $H_{p_2}^2 > 0$ is called **diffractive point** and is denoted by Σ_b^{2-}, the points with $H_{p_2}^2 > 0$ is called **gliding point**, and is denoted by Σ_b^{2+}. All points in $\Sigma_b^{(3)}$ is called **glancing points of higher order**.

Next we indicate that at any glancing point the bicharacteristic strip is tangential to the boundary. Indeed, assume that the operator P has been reduced to the form with the principal symbol $\xi^2 + r(x, y, \eta)$, then $H_p x = 2\xi, H_p^2 x = -2r_x$. Hence $r_x < 0$ at diffractive point. The bicharactristic strip of P is defined by

$$\frac{dx}{ds} = 2\xi,\ \frac{dy}{ds} = r_\eta,\ \frac{d\xi}{ds} = -r_x,\ \frac{d\eta}{ds} = -r_y.$$

Hence at any diffractive point

$$\frac{dx}{ds} = 2\xi = 0,$$

$$\frac{d^2x}{ds^2} = 2\frac{d\xi}{ds} = -2r_x > 0.$$

It means that the bicharacteristic strip arrives at the boundary from the inner part of the domain, and then come back to the inner part. For the gliding point the bicharacteristic strip can only be tangential at the outside of the domain. In other words, in the inner part of the domain there is no gliding bicharacteristic strip tangential to the boundary. At any glancing point of higher order the bicharacteristic strip is tangential to the boundary in higher order.

By using the classification of Σ_b we can define generalized bicharacteristic strip as follows.

Definition 2.4. Generalized bicharacteristic strip is a map γ from $I \setminus B$ to $\Gamma \subset \Sigma_b$, where I is an inteval in R^1, B is a set of isolate points. The map satisfies

(1) If $\gamma(t_0) \in \Sigma_b^0 \cup \Sigma_b^{2-}$, then $\gamma(t) = (x(t), y(t), \xi(t), \eta(t))$ is differentiable at t_0, and $\gamma'(t_0) = H_p(\gamma(t_0))$;

(2) If $\gamma(t_0) \in \Sigma_b^{2+} \cup \Sigma_b^{(3)}$, then the projection $((x(t), y(t), \eta(t))$ of $\gamma(t)$ is differentiable at $t = t_0$, and $x'(t_0) = 0, (y'(t_0), \eta'(t_0)) = H_{r_0}(y(t_0), \eta(t_0))$.

(3) If $t_1 \in B$, then $\gamma(t) \in \Sigma_b^0$ as $|t - t_0| > 0$ small enough. Moreover, $\gamma(t_1 \pm 0)$ exist, which are the different points on a same fiber based on a given point of $\{x = 0\}$.

Based on the above preparations we can describe the singularity of u on the manifold with boundary by using boundary wave front set $WF_b(u)$. For any point, which is not on ∂T^*M, $WF_b(u)$ is nothing but $WF(u)$. For the point on $T^*\partial M$, $WF_b(u)$ is the complimentary set of the points, which are microlocally regular up to the boundary. Here a point $(y_0, \eta_0) \in T^*\partial M$ is "microlocally regular up to the boundary" means that there is a pseudodifferential operator $\psi(y, D_y)$ defined in a neighborhood of (y_0, η_0), such that for some $\epsilon > 0$, $\psi(y, D_y)u(x, y)$ is a $C^\infty([0, \epsilon] \times R^n)$ function.

On the singularity of solutions of the problem (2.71) in the neighborhood of any glancing point the following theorem holds.

Theorem 2.15. *Under the assumptions on the operators P and B, if u is the solution of the problem (2.71), $\rho \in WF_b(u)$, then the generalized bicharacteristic strip $F_\rho(s)$ through ρ belongs to $WF_b(u)$.*

Obviously, when $\rho \in \Sigma_b^0$, the above conclusion is nothing but the conclusion of Theorem 2.8. When $\rho \in \Sigma_b^1$, the conclusion can be derived from Theorem 2.14. When $\rho \in \Sigma_b^{2-}$, R.Melrose and M.Taylor obtained the conclusion of Theorem 2.15 (see [94], [131]). When $\rho \in \Sigma_b^{2+}$ or $\rho \in \Sigma_b^{(3)}$, R.Melrose and J.Sjostrand established the above conclusion (see [99]). Later, L.Hormander gave a unified treatment and proof in [72].

When the order of the operator P is greater than 2, we can use factorization of operators to reduce the problem to the case when the main operator considered is a second order operator. Then the corresponding result can be established by applying Theorem 2.15. The details are omitted here.

The most recent works [100], [101], [136] also discussed the case when the boundary itself has edge or vertex and a bicharacteristics bearing singularity hits this edge or vertex.

2.4.2 *The operators with multiple characteristics*

So far we only discussed the singularity analysis for the operators of principal type. The above discussion does not work if multiple characteristics appear. It means that there is at least a point $(x_0, \xi_0) \in p_m^{-1}(0)$, such that $\nabla_{x,\xi} p_m(x_0, \xi_0) = 0$.

The operators with multiple characteristics can be further classified according to the multiplicity of characteristics. If at any point $(x_0, \xi_0) \in p_m^{-1}(0)$ the multiplicity is constant, then the operator is called **operator with constant multiple characteristics**. Otherwise, it is called **operator with variable multiple characteristics**. Generally, the study on the operators with constant multiple characteristics is less difficulty than the latter.

In the discussion of the singularity analysis for the operators with constant multiple characteristics the operators have to satisfy the **Levi condition**, which is a restriction on the lower order terms. For instance, if P is an operator of second order with multiple characteristics, the symbol of P is

$$p(x, \xi) = q(x, \xi)^2 + p_1(x, \xi),$$

where q and p_1 are the symbols of first order, then Levi condition means that the symbol

$$p_1 - \frac{1}{i} \sum_{j=1}^{n} \frac{\partial q}{\partial x_j} \frac{\partial q}{\partial \xi_j}$$

is of order zero on $q = 0$. Therefore, there is a symbol h of order zero and a corresponding operator H, such that

$$p_1 = h_1 q + \frac{1}{i} \sum_{j=1}^{n} \frac{\partial q}{\partial x_j} \frac{\partial q}{\partial \xi_j}.$$

In accordance

$$P = Q^2 + HQ + \text{operators of order } 0.$$

Generally, if the principal symbol of P has decomposition in a neighborhood of (x_0, ξ_0)

$$p_m = q_1^{r_1} \cdots q_s^{r_s}, \tag{2.73}$$

where each q_j is a symbol of an operator of principal type in strict sense, then the Levi condition is

$$e^{it\phi}P(ae^{it\phi}) = O(t^{m-r_j}) \tag{2.74}$$

holds for $1 \le j \le s$, where ϕ satisfies $d\phi(x_0) = \xi_0$, and $q_j(x, d\phi) = 0$ in a neighborhood of (x_0, ξ_0), $\alpha \in C^\infty$. Then we have the following theorems.

Theorem 2.16. *If P is an operator with constant multiple characteristics satisfying Levi condition, its symbol has the factorization (Eq. (2.73)), u is the real solution of $Pu = f$, then $WF(u) \setminus WF(f) \in p_m^{-1}(0)$, and on $q_j^{-1}(0)$ $WF(u) \setminus WF(f)$ is invariant under the Hamilton flow of H_{q_j}. Therefore, if $f \in C^\infty$, $(x_0, \xi_0) \notin WF(u)$, $q_j(x_0, \xi_0) = 0$, γ is a bicharacteristic strip through (x_0, ξ_0), then $\gamma \cap WF(u) = \emptyset$.*

The corresponding theorem on propagation of singularity of finite order is:

Theorem 2.17. *If P is an operator with constant multiple characteristics satisfying Levi condition, its symbol has the factorization (2.62), u is the real solution of $Pu = f$, then $WF_{s+m-r_j}(u) \setminus WF_s(f) \in p_m^{-1}(0)$. Moreover, $WF_{s+m-r_j}(u) \setminus WF_s(f)$ on $q_j^{-1}(0)$ is invariant under the Hamilton flow of H_{q_j}.*

The proof of the above two theorem can be found in [24]. Moreover, we can also derive the results on singularity reflection for operators with constant multiple characteristics [25].

For the operators with variable multiple characteristics the property of the distribution of singularity is closely related to the manner of the multiplicity of characteristics. Further classification on these operators is necessary in order to give deeper study. Weakly hyperbolic operators form a special class of the operators with variable multiple characteristics. Readers can find more detailed discussions in [61], [63], [69], [111] and the references therein.

As we mentioned in Chapter 1, P.D.Lax [83] indicated that the singularity propagation is essentially related to propagation of waves with high frequency. The authors of [79] continued the study and established many finer results. Because of the limited space of the book we will not discuss more on this topic and prefer to turn to the singularity analysis of nonlinear partial differential equations starting from the next chapters.

Chapter 3

Singularity analysis for semilinear equations

Starting from this chapter we will mainly study singularity analysis for solutions to nonlinear partial differential equations. The phenomena of singularity propagation for nonlinear equations is much more plentiful than that for linear equations. One reason is that the nonlinear functions appearing in the equations often cause interaction of singularity of the solution, another reason is that the characteristics (or bicharacteristics) bearing singularity depends on the solution itself, so that the characteristics itself may also have singularity. Different type of singularities and different type of nonlinearity of equations let the problems be quite many and varied. Like the study on other problems in nonlinear partial differential equations, there is not a unified method to deal with them. Next we will give detailed analysis for some typical cases, and can only give a review or sketchy description for more other cases.

According to the classification in partial differential equations, if the coefficients of the derivatives of unknown functions with highest order in the equation are independent of the unknown function, the equation is called semilinear equation. If the coefficients of the derivatives of unknown functions with highest order in the equation depend on the unknown function and its derivatives of lower order, while the equation is still linear with respect to the derivatives of highest order, then the equation is called quasilinear equation. In more general case the equation is called fully nonlinear equation. This chapter is first devoted to study the singularity analysis for semilinear equation.

3.1 Theorem of propagation of 2s weak singularity

In some cases the result established in Chapter 2 on singularity propagation for linear equations is also valid for semilinear equations with some additional restrictions.

Consider the semilinear wave equation

$$\Box u = f(u), \tag{3.1}$$

where $f(u)$ is a C^∞ function of u. For the solutions to Eq. (3.1) we have the following theorem

Theorem 3.1. *Assume that u is an H^s solution to Eq. (3.1), $s > n/2$, γ is a bicharacteristic strip through (x_0, ξ_0), $s \leq r \leq 2s - n/2$, then $u \in H^s(x_0, \xi_0)$ implies $u \in H^s(\gamma)$.*

The proof of the theorem is based on the following lemma.

Lemma 3.1. *Let n be the dimension of the Euclid space R^n, $f(u)$ be a C^∞ function of u, $n/2 \leq r \leq 2s - n/2$, then $u \in H^r(x_0, \xi_0) \cap H^s$ implies $f(u) \in H^r(x_0, \xi_0)$.*

Proof of Theorem 3.1. Assuming Lemma 3.1 be true we prove Theorem 3.1. According to the assumptions of the theorem, we can find $r_1 \geq s$, such that $u \in H^{r_1}(\gamma)$. Then Lemma 3.1 implies $f(u) \in H^{r_1}(\gamma)$, provided $r_1 \leq 2s - n/2$. Take $r_2 = \min(r_1 + 1, r)$, by using the theorem on singularity propagation of solutions to linear equations (Theorem 2.9) we obtain $u \in H^{r_2}(\gamma) \cap H^s$. It means that we have improved the microlocal regularity of u from H^{r_1} to H^{r_2}. If $r_2 < r$, we can continue this process once again. Then the bootstrap way leads to our conclusion. Theorem 3.1 is thus proved. □

The proof of Theorem 3.1 is not long, but is typical. Its main idea is to decompose the proof on propagation of singularities of solutions to nonlinear equations to two steps. One is the propagation of singularities for linear equations, the other is the composition of singularities by a nonlinear function. When these two steps are well prepared, the result of the theorem on singularity propagation can be obtained immediately. Such a method will frequently be employed later.

Turn to the proof of Lemma 3.1. The lemma means that under the assumptions on the indices r and s the space $H^r(x_0, \xi_0) \cap H^s$ is close with respect to the nonlinear composition. The closeness is essential in the study of singularity analysis for solutions to nonlinear partial differential equations and will also used later. The following several lemmas are the preparation for the proof of Lemma 3.1.

Lemma 3.2 (Schauder). *(1) If $u \in H^s, v \in H^t$ with $s > n/2, 0 \le t \le s$, then $uv \in H^t$;*

(2) If $f(u)$ is a C^∞ function of u, $u(x) \in H^s$, $s > n/2$, then $f(u(x)) \in H^s$.

Proof. The theorem can be proved by different methods. Next we use the theory of paradifferential operators to give a brief proof.

In the case (1), according to the theory of paradifferential operators (see Theorem A.11 in Appendix) the product uv can be written as

$$uv = T_u v + r_1(u, v), \tag{3.2}$$

where T_u is a linear operator from H^t to H^t, $r_1(u, v) \in H^{s+t-n/2} \subset H^t$. Hence $uv \in H^t$.

In the case (2), by using Theorem A.20 in Appendix we can write

$$f(u(x)) = T_{f(u)} u(x) + R(x), \tag{3.3}$$

where $T_{f(u)}$ is a linear operator from H^t to H^t, $R(x) \in H^{2s-n/2} \subset H^t$. Hence the right hand side of Eq. (3.3) belongs to H^s. This is the what we need. □

Lemma 3.3. *Assume that $K(\xi, \eta) : R^n \times R^n \to C^n$ is a local integral function, at least one of the following two inequalities holds:*

$$\sup_\xi \int |K(\xi, \eta)|^2 d\eta < \infty,$$

$$\sup_\eta \int |K(\xi, \eta)|^2 d\xi < \infty, \tag{3.4}$$

then the map

$$(g, h) \mapsto \int K(\xi, \eta) g(\xi - \eta) h(\eta) d\eta$$

is a map from $L^2(R^n) \times L^2(R^n)$ to $L^2(R^n)$. Moreover,

$$\left\| \int K(\xi, \eta) g(\xi - \eta) h(\eta) d\eta \right\|_{L^2} \le C \|g\|_{L^2} \|h\|_{L^2}. \tag{3.5}$$

Proof. Assume that the second inequality in Eq. (3.4) holds, then by using Schwarz inequality we have

$$\left|\int K(\xi,\eta)g(\xi-\eta)h(\eta)d\eta\right|^2$$
$$\leq \int |K(\xi,\eta)h(\eta)|^2 \int |g(\xi-\eta)|^2 d\eta$$
$$\leq \|g\|_{L^2}^2 \int |K(\xi,\eta)h(\eta)|^2 d\eta$$

Hence Fubini theorem implies

$$\left\|\int K(\xi,\eta)g(\xi-\eta)h(\eta)d\eta\right\|_{L^2}^2$$
$$\leq \|g\|_{L^2}^2 \int \left(\int |K(\xi,\eta)h(\eta)|^2 d\xi\right) d\eta$$
$$\leq \|g\|_{L^2}^2 \left(\sup_\eta \int |K(\xi,\eta)|^2 d\xi\right) \|h\|_{L^2}^2,$$

which leads to Eq. (3.5). Similar argument shows that the first inequality in Eq. (3.4) also implies Eq. (3.5). □

Lemma 3.4. *If $u_i \in H^{r_i}(R^n)$, $r_i \leq n/2$ $(i = 1, 2)$, $r_1 + r_2 \geq 0$, then $u_1 u_2 \in H^{(r_1+r_2-n/2)^-}$, where $H^{r^-} = \bigcap_{\epsilon>0} H^{r-\epsilon}$.*

Proof. According to the requirement of the lemma we have to prove $\langle\xi\rangle^{r_1+r_2-n/2-\epsilon}\widehat{u_1u_2}(\xi) \in L^2$ for any $\epsilon > 0$, where $\langle\xi\rangle = (1 + |\xi|^2)^{1/2}$. To this end we define

$$G(\xi,\eta) = \frac{\langle\xi\rangle^{r_1+r_2-n/2-\epsilon}}{\langle\xi-\eta\rangle^{r_1}\langle\eta\rangle^{r_2}},$$

then

$$\langle\xi\rangle^{r_1+r_2-n/2-\epsilon}\widehat{u_1u_2}(\xi) = \int G(\xi,\eta)f(\xi-\eta)g(\eta)d\eta, \tag{3.6}$$

where $f \in L^2, g \in L^2$. According to Lemma 3.3 we only have to prove that $G(\xi,\eta)$ satisfies Eq. (3.3). The fact can be obtained by checking following estimates.

(1) If $|\xi-\eta| \leq |\xi|/2$, then

$$|G(\xi,\eta)| \leq \frac{C}{\langle\xi\rangle^{n/2+\epsilon-r_1}\langle\xi-\eta\rangle^{r_1}} \leq \frac{C}{\langle\xi-\eta\rangle^{n/2+\epsilon}}.$$

(2) If $|\xi-\eta| \geq |\xi|/2$ and $|\eta| \leq |\xi|/2$, which also implies $|\xi-\eta| \leq 3|\xi|/2$, then

$$|G(\xi,\eta)| \leq \frac{C}{\langle\xi\rangle^{n/2+\epsilon-r_2}\langle\eta\rangle^{r_2}} \leq \frac{C}{\langle\eta\rangle^{n/2+\epsilon}}.$$

(3) If $|\xi - \eta| \geq |\xi|/2$ and $|\eta| \geq |\xi|/2$, then

$$|G(\xi, \eta)| \leq \frac{C}{\langle\xi\rangle^{n/2+\epsilon}}$$

can be directly obtained. Summing up, we obtain the lemma. □

Lemma 3.5. *Assume that K_1, K_2, K_3 are cones in R^n_ξ, $u_i \in H^{s_i}(R^n)$, $\Pi_\xi WF(u_i) \in K_i$ hold for $i = 1, 2$, then*

(1) If $K \subset\subset K_2^c$, and $\ell \geq 0, s_1 - \ell \geq 0, s_2 + \ell > \frac{n}{2}$, then $\chi_K(D)(u_1u_2) \in H^{s_1-\ell}(R^n)$.

(2) If $K \subset\subset K_1^c \cap K_2^c$, and $s_1 + s_2 > \frac{n}{2}$, then $\chi_K(D)(u_1u_2) \in H^{s_1+s_2-n/2}(R^n)$.

Here K_i^c is the supplementary set of K_i, χ_K is the characteristic function of the set K.

Proof. Since $\hat{u_i}$ is rapidly decreasing outside of K_i, then we can assume that the support of $\hat{u_i}$ is in K_i without loss of generality. To prove (1) we have to show

$$\langle\xi\rangle^{s_1-k}\chi_K(\xi)\widehat{u_1u_2}(\xi) \in L^2. \tag{3.7}$$

Write the expression in Eq. (3.7) as the form

$$\int G(\xi, \eta)f(\xi - \eta)g(\eta)d\eta,$$

where $f(\xi-\eta) = u_1(\xi-\eta)\langle\xi-\eta\rangle^{s_1}$, $g(\eta) = u_2(\eta)\langle\eta\rangle^{s_2}$, then $G(\xi, \eta)$ satisfies

$$|G(\xi, \eta)| \leq \frac{\langle\xi\rangle^{s_1-k}\chi_K(\xi)\chi_{K_1}(\xi - \eta)\chi_{K_2}(\eta)}{\langle\xi - \eta\rangle^{s_1}\langle\eta\rangle^{s_2}}.$$

Now if $K \subset\subset K_2^c$, then there exists an $\epsilon > 0$, such that

$$\langle\xi - \eta\rangle \geq \epsilon(\langle\xi\rangle + \langle\eta\rangle) > \epsilon\langle\xi\rangle,$$

on the support of G. Hence

$$|G(\xi, \eta)| \leq \frac{C}{\langle\xi - \eta\rangle^{s_1}\langle\eta\rangle^{s_2}}.$$

Therefore, for $s_2 + \ell > \frac{n}{2}$ we have

$$\langle\xi\rangle^{s_1-k}\chi_K(\xi)\widehat{u_1u_2}(\xi) \in L^2.$$

This is the conclusion (1).

In order to prove (2), we apply the trace theorem from R^{2n} to R^n. Then

$$\begin{aligned}
&\|\chi_K(D_x)(u_1u_2)(x)\|_{H^{s_1+s_2-n/2}(R^n)} \\
&= \|[\chi_K(D_x + D_y)u_1(x)u_2(y)]_{y=x}\|_{H^{s_1+s_2-n/2}(R^n)} \\
&\leq \|\chi_K(D_x + D_y)u_1(x)u_2(y)\|_{H^{s_1+s_2}(R^{2n})}.
\end{aligned}$$

The right hand side equals to

$$\|\langle \xi, \eta\rangle^{s_1+s_2}\chi_K(\xi+\eta)\chi_{K_1}(\xi)\hat{u_1}(\xi)\chi_{K_2}(\eta)\hat{u_2}(\eta)\|_{L^2(R^{2n})}.$$

Since $K \subset\subset K_1^c \cap K_2^c$, then in the support of $\chi_K(\xi+\eta)\chi_{K_1}(\xi)\chi_{K_2}(\eta)$ the following inequality holds

$$\langle \xi\rangle \geq \epsilon(\langle \xi+\eta\rangle + \langle \eta\rangle), \quad \langle \eta\rangle \geq \epsilon(\langle \xi+\eta\rangle + \langle \xi\rangle).$$

Hence $\langle \xi\rangle \sim \langle \eta\rangle$, which leads to

$$\chi_K(\xi+\eta)\langle \xi, \eta\rangle^{s_1+s_2}\chi_{K_1}(\xi)\chi_{K_2}(\eta) \leq C\langle \xi\rangle^{s_1}\langle \eta\rangle^{s_2}.$$

Then we have

$$\begin{aligned}
&\|\chi(D_x)(u_1u_2)(x)\|_{H^{s_1+s_2-n/2}(R^n)} \\
&\leq C\|\langle \xi\rangle^{s_1}\hat{u}_1(\xi)\langle \eta\rangle^{s_2}\hat{u}_2(\eta)\|_{L^2(R^{2n})} \\
&= C\|u_1\|_{H^{s_1}(R^n)}\|u_2\|_{H^{s_2}(R^n)}.
\end{aligned}$$

This is the conclusion (2). □

Proof of Lemma 3.1. First from Schauder Lemma (Lemma 3.2) we know that the space $H^{n/2}$ is invariant under nonlinear composition as $s > n/2$. Hence Lemma 3.1 is valid for $r = s$. When $r > s$ we take $\delta = \min(s - n/2, 1)$, and let ρ satisfy $s \leq \rho - \delta, \rho \leq r$. If we can prove that the validity of the lemma for $r = \rho - \delta$ implies the validity of the lemma for $r = \rho$, then the bootstrap way can lead us to the conclusion of Lemma 3.1. Therefore, in the following discussion we may assume that the lemma has been true for $r = \rho - \delta$. By differentiating the function $f(x, u)$ we have

$$D(f(x,u)) = g(x,u) + f'(x,u)Du. \tag{3.8}$$

The first term in the right hand side $g(x,u) \in H^s \cap H^{\rho-\delta}(x_0, \xi_0)$, the second term has the form vDu, where $v \in H^s \cap H^{\rho-\delta}(x_0, \xi_0)$, and $Du \in H^{s-1} \cap H^{\rho-1}(x_0, \xi_0)$. By using the partition of unity we only need to prove that for any sufficiently small cones K_1, K_2 in the space R^n_ξ

$$\chi_K(D)(\chi_{K_1}(D)v \cdot \chi_{K_2}(D)Du) \in H^{\rho-1}(R^n). \tag{3.9}$$

Equation (3.9) can be verified in several cases. If K_1, K_2 are in the neighborhood of ξ_0, then $\chi_{K_1}(D)v \in H^{\rho-1}$, $\chi_{K_2}(D)Du \in H^{\rho-1}$ due to $\delta \leq 1$. In this case Eq. (3.9) can be derived from Schauder Lemma. If K_1 is near to ξ, but K_2 is separated away from K, then Eq. (3.9) can be derived from the conclusion (1) in Lemma 3.5 with $\ell = 1 - \delta$, because of $(s-1) + (1-\epsilon) > n/2$. Similarly, If K_2 is near to ξ, but K_1 is separated away from K, then Eq. (3.9) can be derived from the conclusion (1) of

Lemma 3.5 with $\ell = 0$. Finally, if both K_1 and K_2 are separated away from K, then the conclusion (2) of Lemma 3.5 implies

$$\chi_K(D)(\chi_{K_1}(D)v \cdot \chi_{K_2}(D)Du) \in H^{s+(s-1)-n/2} \subset H^{\rho-1}(R^n),$$

because $\rho \leq r \leq 2s - n/2$. Summing up the above discussion we have $D(f(x,u)) \in H^{\rho-1}(x_0, \xi_0)$, then $f(x,u) \in H^{\rho-1}(x_0, \xi_0)$. If $\rho < r$, we can apply the above argument once and again, and finally establish the conclusion of Lemma 3.1.

The statement of Theorem 3.1 and Theorem 2.2 seems similar. Here we should emphasize two differences. First, the solution in Theorem 3.1 is H^s ($s > n/2$) regular at least. It means that the singularity of solutions could not be too strong. Second, the regularity index of solutions on bicharacteristic strip could not exceed $2s - n/2$. It means that the singularity propagating on bicharacteristic strip could not be too weak. The above two restrictions will always appear in the theorem of singularity propagation for solutions to nonlinear equations, though in many cases the restrictions can be somehow alleviated. Since the restriction $r \leq 2s - n/2$ appears in Theorem 3.1, then the theorem is also called as theorem of propagation of 2s weak singularity.

For more general semilinear wave equation

$$\Box u = f(u, Du), \tag{3.10}$$

the technique applied in the proof of Theorem 3.1 is not available, because of the smoothness loss of order 1 for wave operator. To overcome this difficulty we can take differentiation on the equation to obtain a system of partial differential equations with same principal part for the unknown functions $U = (u, D_{x_1}u, \cdots, D_{x_n}u)$. Then from the conclusion on the propagation of singularities of U to establish the theorem for u.

Theorem 3.2. *Assume that u is an H^s solution to Eq. (3.10), $s > n/2+1$, γ is a bicharacteristic strip passing through (x_0, ξ_0), $s \leq r \leq 2s - n/2 - 1$, then $u \in H^r(x_0, \xi_0)$ implies $u \in H^r(\gamma)$.*

Theorem 3.1 and Theorem 3.2 can be extended to the case for general semilinear hyperbolic equations of higher order.

Theorem 3.3. *Assume that $P = p(x, D_x)$ is a strictly hyperbolic partial differential operator, whose principal symbol $p_m(x,\xi)$ as a polynomial of ξ_n has m real roots different from each other. Assume that u is an H^s solution to*

$$p(x, D)u = f(x, u, \cdots, D^{m-1}u), \tag{3.11}$$

where $s > m + n/2 - 1$, f is a C^∞ function of its arguments. If γ is a bicharacteristic strip through (x_0, ξ_0), $s \leq r \leq 2s - n/2 - m + 1$, then $u \in H^r(x_0, \xi_0)$ implies $u \in H^r(\gamma)$.

By using Theorem 3.3 we can obtain a global regularity theorem for solutions to hyperbolic equations. Consider a hyperbolic equation in $\Omega \subset R^2$. Denote $\Omega_\pm = \Omega \cap \{\pm x > 0\}$, and assume that Ω_+ is included in the domain of determinacy of Ω_-, i.e. any downward bicharacteristics issuing from any point in Ω_+ must enter Ω_- before it leaves Ω, then we have the following theorem.

Theorem 3.4. *Assume that $P = p(x, D_x)$ is a strictly hyperbolic partial differential operator with respect to x_n of order m, u is an H^s solution to*

$$p(x, D)u = f(x, u, \cdots, D^{m-1}u), \tag{3.12}$$

$s > n/2 + m - 1$. Then $u \in H^r(\Omega_-)$ implies $u \in H^r(\Omega_+)$.

Proof. Take a point $x_0 \in \Omega_+$, then consider (x_0, ξ) for all $\xi \in T^*_{x_0}(\Omega)$. If (x_0, ξ) is an elliptic point of $p(x, D_x)$, then $f \in H^s$ implies $u \in H^{s+m}(x_0, \xi)$. If (x_0, ξ) is a characteristic point of $p(x, D_x)$, then we draw the bicharacteristic strip γ issuing from (x_0, ξ). The projection of γ on the base space enter the domain Ω_-. Denote $r_1 = \min(r, s + 1)$, then $u \in H^{r_1}(\Omega_-)$. Since

$$r_1 - m + 1 > s - m + 1 > \frac{n}{2},$$

then the nonlinear composition $f(u) \in H^{r_1 - m + 1}$. According to Theorem 3.3 $u \in H^{r_1}(\gamma)$, which means $u \in H^{r_1}(x_0, \xi)$. Therefore, we have proved that u is microlocal H^{r_1} regular in each direction of $T^*_{x_0}(\Omega)$, then by using the property of wave front set we know $u \in H^{r_1}(x_0)$. In view of the arbitrariness of $x_0 \in \Omega_+$, we know $u \in H^{r_1}(\Omega_+)$. Now if $r_1 < r$, we can take $r_2 = \min(r, r_1 + 1)$ and give the same argument to increase the regularity of u further. Finally we have $u \in H^r(\Omega_+)$. □

Remark 3.1. The conclusion of Theorem 3.4 is also valid for the hyperbolic system with symbol as a diagonal matrix:

$$\begin{pmatrix} P & & \\ & \ddots & \\ & & P \end{pmatrix} \begin{pmatrix} u_1 \\ \vdots \\ u_n \end{pmatrix} + B \begin{pmatrix} u_1 \\ \vdots \\ u_n \end{pmatrix} = \begin{pmatrix} F_1 \\ \vdots \\ F_n \end{pmatrix}, \tag{3.13}$$

where $F_i = F_i(x, u, \cdots, D^\beta u_j)$ is a C^∞ function of the arguments $x, p_{\beta,j}(|\beta| \leq m - 1, 1 \leq j \leq n)$. The proof is similar to Theorem 3.4, and is omitted here.

3.2 Theorem on propagation of 3s weak singularity

In the discussion of Section 3.1 we find that Lemma 3.1 plays the key role in the proof of the conclusion on 2s singularity propagation. For the strictly hyperbolic equations of second order by using the convexity of the characteristic cone we can introduce a suitable microlocal space, which can describe propagation of weaker singularities.

Theorem 3.5. *Assume that u is an H^s solution to*

$$\Box u = f(x, u), \tag{3.14}$$

$s > n/2$, γ is a bicharacteristic strip through (x_0, ξ_0), $s \le r \le 3s - n + 1$, then $u \in H^r(x_0, \xi_0)$ implies $u \in H^r(\gamma)$.

The theorem is also called the theorem of propagation of 3s singularities. In order to prove it we need to introduce a new functional space to describe the special regularity of the solutions to Eq. (3.14). Besides, we emphasize that since $u \in H^s \cap H^r(x_0, \xi_0)$ is a solution of a wave equation, it can have some additional regularity. By using such a regularity related to wave operator we can alleviate the restriction to u in Theorem 3.1.

Definition 3.1. If $\Box^j u \in H^{s-j}$ for any j satisfying $0 \le j \le s - n/2$, and $\Box^k u \in H^{(2s-n/2-2k)-}$ for $s - n/2 \le k < s - n/2 + 1$, then we say $u \in \tilde{H}^s(\Box)$.

It is not difficult to prove that the space $\tilde{H}^s(\Box)$ is close under nonlinear composition by using the chain rule, Leibnitz Formula and Schauder Lemma for Sobolev space. Indeed, we have the following lemma.

Lemma 3.6. *If $u \in \tilde{H}^s(\Box)$,$s > n/2$, $f(x, u)$ is a C^∞ function of its arguments, then $f(x, u(x)) \in \tilde{H}^s(\Box)$.*

Proof. Leibnitz Formula in calculus implies

$$\Box^j f(x, u) = \tilde{f}(x, u, \cdots, \Box^{\alpha_i} D^{\alpha_i'} u, \cdots), \quad \alpha_i + |\alpha_i'| \le j. \tag{3.15}$$

If $j < s - n/2$, then $\Box^{\alpha_i} D^{\alpha_i'} u \in H^{s-j}$. Hence $u \in \tilde{H}^s(\Box)$.

If $s - n/2 \le k < s - n/2 + 1$, then

$$\begin{aligned} \Box^k f(x, u) &= \Box f(x, u, \cdots, \Box^{\alpha_i} D^{\alpha_i'} u, \cdots) \\ &= \sum_i \tilde{f}_i \cdot \Box^{\alpha_i + 1} D^{\alpha_i'} u + \sum_{i,j,\ell} \tilde{f}_{ij} \cdot \Box^{\alpha_i} D^{\alpha_i'} D_\ell u \cdot \Box^{\alpha_j} D^{\alpha_j'} D_\ell u. \end{aligned} \tag{3.16}$$

Since $\tilde{f}_i \in H^{s-k+1}$ and $\Box^{\alpha_i+1} D^{\alpha_i'} u \in H_-^{2s-n/2-2k}$, then the first term in the right hand side of Eq. (3.16) belongs to $H_-^{2s-n/2-2k}$. The second term takes the form $H^{s-k+1} \cdot H^{s-k} \cdot H^{s-k}$, which is contained in $H^{s-k} \cdot H^{s-k}$. Since $2s - 2k > n - 2 \geq 0$, then Lemma 3.2 implies that the term is in $H^{(2s-n/2-2k)_-}$. □

Next we consider the closeness of the space with assigned microlocal regularity under nonlinear composition.

Lemma 3.7. *If* $u \in \tilde{H}^s(\Box) \cap H^g(x_0, \xi_0)$,$n/2 < s \leq g < 3s - n$, $f(x, u)$ *is a* C^∞ *function of its arguments, then* $f(x, u(x)) \in \tilde{H}^s(\Box) \cap H^g(x_0, \xi_0)$.

Lemma 3.7 is in fact the key to establish the result in this section. To prove it we have to do some preparations. Before we prove it let us first indicate how this Lemma leads to the conclusion of Theorem 3.5.

Proof of Theorem 3.5. Based on Lemma 3.7 the conclusion in Theorem 3.5 can be proved by using bootstrap way.

First, we prove $u \in \tilde{H}^s(\Box)$. Indeed, when $j < s - n/2$,

$$\Box^j u = \tilde{f}(x, u, \cdots, D^j u). \tag{3.17}$$

Schauder Lemma implies $\tilde{f}(x, u, \cdots, D^j u) \in H^{s-j}$, then $\Box^j u \in H^{k-j}$. Differentiating Eq. (3.17) we have

$$\begin{aligned}\Box^k u = h(x, u, \cdots, D^{k-1}u) + \sum_{|\alpha|=k} h_\alpha(x, u, \cdots, D^{k-1}u) D^\alpha u \\ + \sum_{|\alpha|=k, |\beta|=k} h_{\alpha\beta}(x, u, \cdots, D^{k-1}u) D^\alpha u D^\beta u.\end{aligned} \tag{3.18}$$

Schauder Lemma implies $H^r \cdot H^\rho \subset H^\rho$ as $|\rho| \leq r, r > n/2$. Therefore, from $2s - n/2 - 2k \leq s - k \leq n/2 < s - (k-1)$ we know that the second term in Eq. (3.18) belongs to $H^{(2s-n/2-2k)_-}$. Similarly, the third term in Eq. (3.18) also belongs to this space by virtue of $2s - 2k \geq n - 2 \geq 0$.

Now we may assume $u \in H^{r_1}(\gamma)$. If $r_1 < 3s - n$, then Lemma 3.7 implies $f(x, u) \in H^{r_1}(\gamma)$. The theorem of singularity propagation for linear equations gives $u \in H^{\min(g, r_1+1)}$. Doing in this way once and again we obtain the conclusion of the theorem. □

In order to prove Lemma 3.7 we should use the convexity of the characteristic cone of the wave operator. To do it we rewrite the wave operator as

$$\Box' = D_{x_1} D_{x_n} - \frac{1}{2} \sum_{j=2}^{n-1} D^2_{x_i x_j}. \tag{3.19}$$

In fact, by a simple transformation of independent variables the operator $\square$ can be transformed to the form (3.19). Let $p(\xi) = Sym(\square')$, we have

$$\xi_n + \eta_n = \frac{p(\xi)}{\xi_1} + \frac{p(\eta)}{\eta_1} + \frac{1}{2}\sum_{j=2}^{n-1} \xi_j^2 \frac{\xi_1 + \eta_1}{\xi_1 \eta_1} - \frac{1}{2}\sum_{j=2}^{n-1} \frac{\xi_j^2 - \eta_j^2}{\eta_1}. \tag{3.20}$$

According to the expression (3.20) we have the following propositions.

Lemma 3.8. *If K is a cone, which does not intersect $\{\xi_n = 0\}$ except the origin. The support of $\hat{u}$ and $\hat{v}$ is in a neighborhood of $(1, 0, \cdots, 0)$ and $(-1, 0, \cdots, 0)$ respectively, then for any $\sigma = \sigma_i + \sigma_j$ we have*

$$\|\chi_K(D_x)uv\|_{H^{\sigma+k-n/2}} \leq \sum_{i+j\leq k} \|(\square')^i u\|_{H^{\sigma_i - i}} \|(\square')^j u\|_{H^{\sigma_j - j}} \tag{3.21}$$

Proof. We will only give the proof for $k = 1$, as for the case $k > 1$ the proof is similar. Write $\chi_K(D_x)D_{x_n}uv(x)$ as

$$\chi_K(D_x + D_y)(D_{x_n} + D_{y_n})\chi_+(D_x)\chi_-(D_y)u(x)v(y)\big|_{y=x},$$

where $\chi_\pm \in S^0_{1,0}(R^n)$, which is equal to 1 in the support of $\hat{u}, \hat{v}$ respectively, and is supported in a conical neighborhood of $(\pm 1, 0, \cdots, 0)$. As indicated in Lemma 3.5 $\langle\xi\rangle \sim \langle\eta\rangle$, then there is a C^∞ function χ_0 supported in $(0, \infty)$, such that

$$\chi_K(\xi+\eta)\chi_+(\xi)\chi_-(\eta) = \chi_K(\xi+\eta)\chi\left(\frac{\langle\xi\rangle}{\langle\eta\rangle}\right)\chi_+(\xi)\chi_-(\eta).$$

By using Eq. (3.20) we have

$$\begin{aligned}&\chi_K(\xi+\eta)(\xi_n+\eta_n)\chi_+(\xi)\chi_-(\eta)\\ &= \chi_K(\xi+\eta)\left(\frac{p(\xi)}{\xi_1} + \frac{p(\eta)}{\eta_1} + \sum(\xi_j+\eta_j)b_j(\xi,\eta)\right)\chi_+(\xi)\chi_-(\eta), \end{aligned}\tag{3.22}$$

where $b_j(\xi,\eta)$ is a $S^0_{1,0}(R^{2n})$ symbol. Besides, all $b_j(\xi,\eta)$ are polynomials of ξ_i/η_1 and η_i/η_1 with $i \neq 1$, which become small as the support of $\chi_\pm$ shrinks. Meanwhile, $(\xi_j+\eta_j)(\xi_n+\eta_n)^{-1}$ is bounded on the support of $\chi_K(\xi+\eta)$, then removing the last term in Eq. (3.22) to the left hand side we obtain

$$\begin{aligned}&\chi_K(\xi+\eta)(\xi_n+\eta_n)\chi_+(\xi)\chi_-(\eta)(1 - \sum(\xi_j+\eta_j)(\xi_n+\eta_n)^{-1}b_j(\xi,\eta))\\ &= \chi_K(\xi+\eta)\left(\frac{p(\xi)}{\xi_1} + \frac{p(\eta)}{\eta_1}\right).\end{aligned}$$

Then

$$\chi_K(\xi+\eta)(\xi_n+\eta_n)\chi_+(\xi)\chi_-(\eta) = \alpha_0(\xi,\eta)\left(\frac{p(\xi)}{\xi_1} + \frac{p(\eta)}{\eta_1}\right),$$

where

$$\alpha_0(\xi,\eta) = \frac{\chi_K(\xi+\eta)\tilde{\chi}_+(\xi)\tilde{\chi}_-(\eta)}{1-\sum(\xi_j+\eta_j)(\xi_n+\eta_n)^{-1}b_j(\xi,\eta)},$$

which is a symbol in the class $S^0_{1,0}(R^{2n})$.

Since D_{x_n} is elliptic on $supp\chi_K$, then

$$\begin{aligned}
&\|\chi_K(D)uv\|_{H^{\sigma-n/2+1}} \le \|\chi_K(D)D_{x_n}uv\|_{H^{\sigma-n/2}(R^n)} \\
&\le C\|p(D_x)D_{x_1}^{-1}\tilde{\chi}(D_x,D_y)u(x)v(y)\|_{H^\sigma(R^{2n})} \\
&\quad +\|p(D_y)D_{y_1}^{-1}\tilde{\chi}(D_x,D_y)u(x)v(y)\|_{H^\sigma(R^{2n})}.
\end{aligned}$$

Notice that $\langle\xi\rangle \sim \langle\eta\rangle$ on the support of $\tilde{\chi}$, then the quantity can also be dominated by

$$C(\|p(D_x)u\|_{H^{\sigma_1-1}(R^n)}\|v\|_{H^{\sigma_2}(R^n)} + \|u\|_{H^{\sigma_1}(R^n)}\|p(D_x)v\|_{H^{\sigma_2-1}(R^n)}),$$

where $\sigma_{1,2}, \rho_{1,2}$ are any choice satisfying $\sigma_1+\sigma_2=\sigma, \rho_1+\rho_2=\rho$. Hence we obtain the conclusion of the lemma in the case $k=1$. As we said the proof for the case $k>1$ can be done in similar way. □

Based on Lemma 3.8 we turn to the 3s microlocal regularity of the product of two functions.

Lemma 3.9. *Assume $n/2 < s \le g < 3s-n$, the characteristics of $\Box$ vanishes at (x_0,ξ_0). If $u,v \in \tilde{H}^s(\Box)\cap H^g(x_0,\xi_0)$, then $vD^\alpha u \in H^{g-|\alpha|}(x_0,\xi_0)$ as $|\alpha|\le 1$. Particularly, $\tilde{H}(\Box)\cap H^{g-|\alpha|}(x_0,\xi_0)$ is an algebra.*

Proof. In the sequel we may assume $g>s$, otherwise the conclusion of the theorem is trivial. We assume that the support of u,v locates is a small neighborhood of x_0, then by using partition of unity in the dual space we can also assume the support of $\hat{u},\hat{v}$ locates in a sufficiently small cone K_1,K_2 respectively. For the conical neighborhood K of ξ_0 we consider the following three cases.

(1) At least one of the supports of K_1,K_2 in the neighborhood of ξ_0:

If both K_1 and K_2 are in a small neighborhood of ξ_0, then $v\in H^g, D^\alpha u\in H^{g-|\alpha|}$. Then Schauder Lemma implies $vD^\alpha u\in H^{g-|\alpha|}$ immediately.

If K_1 in a small neighborhood of ξ_0, K_2 does not intersect K, then $v\in H^g, D^\alpha u\in H^{s-|\alpha|}$. In view of $(s-|\alpha|)+|\alpha|>\frac{n}{2}$, then the conclusion (1) of Lemma 3.5 implies $\chi_K(D)(vD^\alpha u)\in H^{g-|\alpha|}$.

If K_2 in a small neighborhood of ξ_0, K_1 does not intersect K, then $v\in H^s, D^\alpha u\in H^{g-|\alpha|}$. The conclusion (1) of Lemma 3.5 also implies $\chi_K(D)(vD^\alpha u)\in H^{g-|\alpha|}$.

(2) If K_1 (K_2 resp.) is away from the characteristic set of $\Box'$, then $\Box'$ is elliptic in K_1. Hence $(\Box')^k v \in H^{2s-n/2-2k-\epsilon}$ implies $v \in H^{2s-n/2-\epsilon}$. Correspondingly, $D^\alpha u \in H^{2s-n/2-|\alpha|-\epsilon}$. Then the conclusion (2) of Lemma 3.5 gives $\chi_K(D)(vD^\alpha u) \in H^{3s-n-|\alpha|-\epsilon}$.

(3) The remains is the case when K_1, K_2 are in a small neighborhood of ξ_1, ξ_2 respectively. Moreover, $\xi_1 \neq \xi_0$, $\xi_2 \neq \xi_0$.

If $\xi_1 \neq -\xi_2$, then Theorem 2.3 implies that $\Pi_\xi WF(vD^\alpha u)$ is included in $E = \Pi_\xi WF(v) + \Pi_\xi WF(D^\alpha u)$. From the convexity of the characteristic set of the operator $\Box'$ we know that E will not meet the characteristic directions other than the directions in the neighborhood of ξ_1 or ξ_2. Hence $E \cap K = \chi_K(D)(vD^\alpha u) \in H^\infty$, provided K is in a small neighborhood of ξ_0.

When $\xi_1 = -\xi_2$, we can apply Lemma 3.8. Letting k be the integer in $(s - n/2, s - n/2 + 1)$ and writing $(3s - n - |\alpha|)-$ as $\sigma + k - n/2$ we have

$$\begin{aligned}\sigma &= \frac{n}{2} - k + (3s - n - |\alpha|) - \epsilon \\ &\geq -\left(s - \frac{n}{2} + 1\right) + \left(3s - \frac{n}{2} - 1\right) - \epsilon \\ &= (2s - 2) - \epsilon \quad > 0.\end{aligned}$$

Then by using Lemma 3.8 we obtain

$$\|\chi_K(D)vD^\alpha u\|_{H^{(3s-n-|\alpha|)-}} \leq C \sum_{i+j\leq k} \|(\Box')^i v\|_{H^{\sigma_i - i}} \|(\Box')^j D^\alpha u\|_{H^{\sigma_j - j}},$$

where $\sigma_i + \sigma_j \geq \sigma$. The values of σ_i and σ_j can be chosen as follows:

$$\sigma_i = s,\ \sigma_j = s - |\alpha|, \quad \text{if} \quad i \neq k, j \neq k;$$

$$\sigma_i = \left(2s - \frac{n}{2} - k\right)-,\ \sigma_j = s - |\alpha|, \quad \text{if} \quad i = k, j = 0;$$

$$\sigma_i = s,\ \sigma_j = \left(2s - \frac{n}{2} - k - |\alpha|\right)-, \quad \text{if} \quad i = 0, j = k;$$

In all three cases $\|(\Box')^i v\|_{H^{\sigma_i - i}}, \|(\Box')^j D^\alpha u\|_{H^{\sigma_j - j}}$ are bounded, then we know $vD^\alpha u \in H^{g-|\sigma|}(x_0, \xi_0)$ as $g < 3s - n$.

Combining the analysis in all cases we establish the lemma. $\Box$

Lemma 3.9 is the special case of Lemma 3.7, while it can also lead us to Lemma 3.7.

Proof of Lemma 3.7. The proof of this lemma is quite similar to that for Lemma 3.1. When $g = s$, the conclusion is given by Lemma 3.1. When $g > s$, by taking $\delta = \min(2s - n, 1)$ we can derive $f(x, u) \in H^\rho(x_0, \xi_0)$ from

$f(x,u) \in H^{\rho-\delta}(x_0,\xi_0)$, where ρ satisfies $s \le \rho-\delta$ and $\rho \le g$. Differentiating $f(x,u)$ gives

$$D_x(f(x,u)) = h(x,u) + f'(x,u)D_x u.$$

By using the assumptions the first term belongs to $H^{\rho-\delta}(x_0,\xi_0) \subset H^{\rho-1}(x_0,\xi_0)$, the second term has the form vDu with $v \in \tilde{H}^s(\Box) \cap H^{\rho-\delta}(x_0,\xi_0)$, $u \in \tilde{H}^s(\Box) \cap H^{\rho}(x_0,\xi_0)$. Then by the same method as in Lemma 3.1 we can derive $f'(x,u)D_x u \in H^{\rho-1}(x_0,\xi_0)$ from Lemma 3.6, provided $\rho < 3s - n$. This means that $D_x(f(x,u)) \in H^{\rho-1}(x_0,\xi_0)$, i.e. $f(x,u) \in H^{\rho}(x_0,\xi_0)$.

As did before, the bootstrap way gives us the conclusion of Lemma 3.7. □

Remark 3.2. The theorem of $3s$ weak singularity propagation is first proved by M.Beals [7], where the conclusion in Theorem 3.5 can also be extended to general semilinear strictly hyperbolic equation of second order. Later, in [89] the conclusion was improved to $r < 3s - n + 2$.

Remark 3.3. For general hyperbolic equation of higher order the theorem of $3s$ weak singularity propagation is not valid. The reason is that the characteristic cone for general hyperbolic equations of higher order is not always convex, while the convexity played a crucial role in our above proof. In [8] a counterexample is given.

3.3 Singularity interaction and singularity index

In the above two sections we proved the conclusion that for the solution to semilinear partial differential equations, under some assumptions its $2s$ or $3s$ weak singularity propagates like that for the solutions to linear equations. How does weaker singularity propagate? The further study found when two characteristics bearing singularity intersect, the intersection may become a source producing new singularity. Such a phenomenon is called **singularity interaction**.

Let us first look an example. Consider a Cauchy problem of system of hyperbolic equations in (t,x) space

$$\begin{cases} (\partial_t + \partial_x)u = 0, \\ (\partial_t - \partial_x)v = 0, \\ \partial_t w = uv. \end{cases} \tag{3.23}$$

satisfying initial data on $t = 0$

$$u = \begin{cases} (x-1)^n, & x > 1, \\ 0, & x \le 1, \end{cases} \qquad v = \begin{cases} 0, & x > -1, \\ (-x-1)^n, & x \le -1, \end{cases} \qquad w = 0. \tag{3.24}$$

Its solution can be written explicitly

$$u = \begin{cases} (x+t-1)^n, & x+t > 1, \\ 0, & x+t \le 1, \end{cases}$$
$$v = \begin{cases} 0, & x-t > -1, \\ (-x+t-1)^n, & x-t \le -1, \end{cases}$$
$$w = \begin{cases} 0, & (t-1)^2 < x^2, \\ \int_{1+|x|}^{t} ((\tau-1)^2 - x^2)^n d\tau, & (t-1)^2 > x^2,\ t > 1. \end{cases} \tag{3.25}$$

Obviously, u is C^∞ everywhere except on $L_1 : x+t = 1$, and belongs to $H^{n+1/2-\epsilon}$ on L_1 with $\epsilon > 0$. v is C^∞ everywhere except on $L_2 : x-t = 1$, and belongs to $H^{n+1/2-\epsilon}$ on L_2 with $\epsilon > 0$. w has singularity on $x \pm t = 1$ and on $L_3 : x = 0, t > 1$. It belongs to $H^{2n+3/2-\epsilon}$ there. Therefore, the whole solution $U = (u, v, w)$ to the problem (3.22), (3.24), it has singularity on L_1 and L_2 as $t < 1$, and has singularity on L_1, L_2 and L_3. The singularity on L_3 is of order $2n + 3/2$, which does not propagate from L_1 or L_2. It comes from a new source at $L_1 \cap L_2$.

The example indicates when two bicharacteristics intersect, the intersection may form a new source of singularity. This is the phenomenon of singularity interaction in nonlinear equations. Due to the interaction of singularities the study on singularity analysis for nonlinear equations is more complicated than that for linear equations. J.Rauch and M.Reed gave a detailed analysis on singularity interaction for the solutions to semilinear hyperbolic systems with one space variable. They also defined the singularity index for solutions to such systems, which can describe the distribution and strength of all singularities for a given solution [116]. Next we are going to discuss the nonlinear wave equation with multiple space variables. A corresponding regularity index will also be defined, which can control all singularities produced by singularity interaction.

Consider the Cauchy problem

$$\Box u \equiv \left(\partial_{x_n}^2 - \sum_{i=1}^{n-1} \partial_{x_i}^2 \right) u = f(x, u), \tag{3.26}$$

$$u|_{x_n=0} = \phi_0, \quad \frac{\partial u}{\partial x_n}|_{x_n=0} = \phi_1. \tag{3.27}$$

Assume that the H^s solution with $(s > n/2)$ to the Cauchy problem exists, we want to determine the distribution and strength of its singularities. In what follows we use wave front set $WF(u)$ or regularity function $s_u(x, \xi)$ to describe singularity of u. Besides, to describe all singularities produced by interaction some new concepts should be introduced.

Definition 3.2. Let $r(x, \xi)$ be a lower semi-continuous function, defined on the cotangent bundle $T^*(\Omega)$, positively homogeneous of degree 0 with respect to ξ. If $r(x, \xi)$ satisfies the following conditions:

(1) $r(x, \xi) > n/2$,

(2) $r(x, \xi) \le r(x, \xi_1) + r(x, \xi_2) - n/2$, where ξ, ξ_1, ξ_2 are three points on S^{n-1}, and ξ is located on the great circle arc connecting ξ_1 and ξ_2, then $r(x, \xi)$ is called **admissible function.**

Lemma 3.10. *If $r(x, \xi)$ is an admissible function on $T^*(\Omega)$ and u, v are distributions on Ω, the regularity functions of u, v satisfy $s_u(x, \xi) \ge r(x, \xi)$, $s_v(x, \xi) \ge r(x, \xi)$, then the inequality $s_{uv}(x, \xi) \ge r(x, \xi)$ holds.*

Proof. Since the regularity function describes the microlocal property, we only need to consider the function in a neighborhood of the given point x. Briefly write the functions $s_u(x, \xi), s_v(x, \xi), r(x, \xi)$ by $s_u(\xi), s_v(\xi), r(\xi)$. From the definition of regularity function we know that for fixed $(\bar{x}, \bar{\xi})$ and $\epsilon > 0$ there is a neighborhood ω of $\bar{x}$ and a conical neighborhood V of $\bar{\xi}$, such that for any $\phi \in C_0^\infty(\omega)$

$$\widehat{\phi u}(\xi)\langle\xi\rangle^{r(\bar{\xi})-\epsilon} \in L^2(V), \tag{3.28}$$

where $\langle\xi\rangle = (1 + |\xi|^2)^{1/2}$. For v we have the same estimate. Next for the notational simplicity we assume that the support of u, v is included in ω, so that the factor ϕ can be omitted.

According to the property of Fourier transformation

$$\widehat{uv}(\xi) = \int \hat{u}(\xi - \eta)\hat{v}(\eta)d\eta. \tag{3.29}$$

For any $\epsilon > 0$ we have to prove

$$\langle\xi\rangle^{r(\xi_0)-\epsilon}\widehat{uv}(\xi) \in L^2 \tag{3.30}$$

in a neighborhood Γ_{ξ_0} of ξ_0. Take $\epsilon > 0$ be sufficiently small, such that $r(\xi) > n/2 + \epsilon$ holds for any ξ. Since $r(\xi)$ is lower semi-continuous, then we can find conical neighborhood Γ_1, Γ_2 such that $\Gamma_1 + \Gamma_1 \subset \Gamma_2$ and

$$r(\xi) > r(\xi_0) - \frac{\epsilon}{2}. \tag{3.31}$$

Consider the integral

$$
\begin{aligned}
&|\langle\xi\rangle^{r(\xi_0)-\epsilon}\chi_{\Gamma_1}(\xi)\widehat{uv}(\xi)| \\
&\le \int |\chi_{\Gamma_1}(\xi)\chi_{\Gamma_2}(\eta)\chi_{\Gamma_2}(\xi-\eta)\langle\xi\rangle^{r(\xi_0)-\epsilon}\hat{u}(\xi-\eta)\hat{v}(\eta)|d\eta \\
&\quad + \int |\chi_{\Gamma_1}(\xi)\chi_{\Gamma_2}(\eta)\chi_{S^{n-1}\setminus\Gamma_2}(\xi-\eta)\langle\xi\rangle^{r(\xi_0)-\epsilon}\hat{u}(\xi-\eta)\hat{v}(\eta)|d\eta \\
&\quad + \int |\chi_{\Gamma_1}(\xi)\chi_{S^{n-1}\setminus\Gamma_2}(\eta)\chi_{\Gamma_2}(\xi-\eta)\langle\xi\rangle^{r(\xi_0)-\epsilon}\hat{u}(\xi-\eta)\hat{v}(\eta)|d\eta \\
&\quad + \int |\chi_{\Gamma_1}(\xi)\chi_{S^{n-1}\setminus\Gamma_2}(\eta)\chi_{S^{n-2}\setminus\Gamma_2}(\xi-\eta)\langle\xi\rangle^{r(\xi_0)-\epsilon}\hat{u}(\xi-\eta)\hat{v}(\eta)|d\eta \\
&= I_1 + I_2 + I_3 + I_4,
\end{aligned}
\tag{3.32}
$$

where $\chi_{\Gamma_1}, \chi_{\Gamma_2}$ are characteristic functions of corresponding sets. Next we use Lemma 3.3 to prove all integral I_k are L^2 integrable. Let

$$
\begin{aligned}
K_1(\xi,\eta) &= \chi_{\Gamma_1}(\xi)\chi_{\Gamma_2}(\eta)\chi_{\Gamma_2}(\xi-\eta)\langle\xi\rangle^{r(\xi_0)-\epsilon}\langle\xi-\eta\rangle^{-r(\xi_0)+\epsilon/2}\langle\eta\rangle^{-r(\xi_0)+\epsilon/2}, \\
K_2(\xi,\eta) &= \chi_{\Gamma_1}(\xi)\chi_{\Gamma_2}(\eta)\chi_{S^{n-1}\setminus\Gamma_2}(\xi-\eta) \\
&\quad \times\langle\xi\rangle^{r(\xi_0)-\epsilon}\langle\xi-\eta\rangle^{-n/2-\epsilon}\langle\eta\rangle^{-r(\xi_0)+\epsilon/2}, \\
K_3(\xi,\eta) &= \chi_{\Gamma_1}(\xi)\chi_{S^{n-1}\setminus\Gamma_2}(\eta)\chi_{S^{n-1}\setminus\Gamma_2}(\xi-\eta) \\
&\quad \times\langle\xi\rangle^{r(\xi_0)-\epsilon}\langle\xi-\eta\rangle^{-r(\xi_0)+\epsilon/2}\langle\eta\rangle^{-n/2-\epsilon/2}, \\
K_4(\xi,\eta) &= \chi_{\Gamma_1}(\xi)\chi_{S^{n-1}\setminus\Gamma_2}(\eta)\chi_{S^{n-1}\setminus\Gamma_2}(\xi-\eta) \\
&\quad \times\langle\xi\rangle^{r(\xi_0)-\epsilon}\langle\xi-\eta\rangle^{-r(\xi_0)+\epsilon/6}\langle\eta\rangle^{-r(\xi_0)+\epsilon/6}.
\end{aligned}
$$

For K_1 , due to

$$\langle\xi\rangle^{r(\xi_0)-\epsilon} \le \langle\xi-\eta\rangle^{r(\xi_0)-\epsilon} + \langle\eta\rangle^{r(\xi_0)-\epsilon}$$

and $r(\xi_0) > n/2$, K_1 satisfies the condition in Lemma 3.3. Hence I_1 is L^2 integrable.

For K_2, where the inequality

$$|\eta| \ge c(|\xi-\eta| + \eta|) \ge c|\xi|$$

holds, then

$$|K_2(\xi,\eta)| \le C\langle\xi\rangle^{r(\xi_0)-\epsilon}\langle\xi-\eta\rangle^{-n/2-\epsilon}\langle\xi\rangle^{-r(\xi_0)+\epsilon/2} \le C\langle\xi-\eta\rangle^{-n/2-\epsilon}.$$

Then K_2 also satisfies the condition of Schauder Lemma. The same discussion can be proceeded for K_3. Hence I_2, I_3 are L^2 integrable.

Finally we consider the integral I_4. Denote $\zeta = \xi - \eta$ and regard η, ζ as points moving on $S^{n-1} \setminus \Gamma_2$. When $(\bar{\xi}, \bar{\eta})$ is in the support of K_1, then

$\bar{\zeta} + \bar{\eta} = \bar{\xi} \in \Gamma_1$, then by using the property of the admissible function we know

$$r(\bar{\eta}) + r(\bar{\zeta}) \geq r(\bar{\xi}) + \frac{n}{2}. \tag{3.33}$$

Besides, we can find small conical neighborhood $C'_{\bar{\eta}}, C''_{\bar{\zeta}}$, such that

$$\hat{v}(\eta)\langle\eta\rangle^{r(\bar{\eta})-\epsilon/6} \in L^2(C'_{\bar{\eta}}), \quad \hat{u}(\eta)\langle\zeta\rangle^{r(\bar{\zeta})-\epsilon/6} \in L^2(C''_{\bar{\zeta}}). \tag{3.34}$$

All couples of conical neighborhood $\{C'_{\bar{\eta}} \times C''_{\bar{\zeta}}\}$ form an open covering of $(S^{n-1}\backslash\Gamma_2)\times(S^{n-1}\backslash\Gamma_2)$, then we can find a finite open covering $\{C'_i \times C''_j\}$, and get a partition of unity $1 = \sum_{ij}\phi_{ij}$ subordinated to this covering. Hence we only have to consider I_4^{ij} corresponding to the kernel $\phi_{ij}K_4$.

Due to the character of the support of ϕ_{ij}, we only need to check whether the function

$$K_4^{ij}(\xi,\eta) = \chi_{C''_i}(\eta)\chi_{C''_j}\chi_{\Gamma_1}(\xi)\chi_{S^{n-1}\backslash\Gamma_2}\chi_{S^{n-1}\backslash\Gamma_2}(\xi-\eta)$$
$$\times\langle\xi\rangle^{r(\xi_0)-\epsilon}\langle\eta\rangle^{-r(\eta_i)+\epsilon/6}\langle\xi-\eta\rangle^{-r(\zeta_j)+\epsilon/6}$$

satisfies the condition of Theorem 3.3, where η_i, ζ_j are corresponding $\bar{\eta}$,$\bar{\zeta}$ in C'_i, C''_j. By using (3.31),(3.33) we have

$$r(\eta_i) - \frac{\epsilon}{6} + r(\zeta_j) - \frac{\epsilon}{6} \geq r(\bar{\xi}) + \frac{n}{2} - \frac{\epsilon}{3} > r(\xi_0) - \epsilon + \frac{n}{2} + \frac{\epsilon}{6}. \tag{3.35}$$

Notice that on the support of K_4^{ij}

$$|\xi - \eta| \geq c(|\xi| + |\eta|) \geq c|\xi|,$$

$$|\eta| \geq c(|\xi - \eta| + |\xi|) \geq c|\xi|,$$

then by using Lemma 3.3 we know I_4^{ij} is L^2 integrable. Hence $I_4 = \sum I_4^{ij}$ is also L^2 integrable, so that (3.30) is proved. Therefore,

$$s_{uv}(\xi_0) \geq r(\xi_0) - \epsilon.$$

It means that $s_{uv}(\xi_0) \geq r(\xi_0)$, because ϵ can be arbitrarily small. Since ξ_0 is an arbitrary point in R^n_ξ, then we are led to the conclusion of Lemma 3.10. □

Definition 3.3. If $g(x,\xi)$ is a given lower semi-continuous function on $T^*(\Omega)$, homogeneous of degree zero with respect to ξ and satisfying $g(x,\xi) > n/2$, then the supremum of all admissible functions, which is not larger than $g(x,\xi)$, is called **melting function** of g and denoted by

$$A[g] = \sup\{r(x,\xi); r \leq g, r \text{ is admissible}\}. \tag{3.36}$$

Lemma 3.11. *Assume that $g(x,\xi)$ is a lower semi-continuous function, homogeneous of degree zero with respect to ξ and satisfying $g(x,\xi) > n/2$. If the distributions u, v satisfy*

$$s_u(x,\xi) \geq g(x,\xi), \quad s_v(x,\xi) \geq g(x,\xi), \tag{3.37}$$

then

$$s_{uv}(x,\xi) \geq A[g](x,\xi). \tag{3.38}$$

Proof. $A[g]$ is also an admissible function. Indeed, since the supremum of a lower semi-continuous function is also lower semi-continuous, then $A[g]$ is also semi-continuous. Now assume that ξ, ξ_1, ξ_2 are three points on S^{n-1}, $\xi \in \widehat{\xi_1\xi_2}$, then the definition of $A[g]$ implies that for any x and $\delta > 0$ there is an admissible function $r^* \leq g$, such that $A[g](x,\xi) < r^*(x,\xi) + \delta$. Hence

$$\begin{aligned} A[g](x,\xi) &< r^*(x,\xi_1) + r^*(x,\xi_2) + \delta - \frac{n}{2} \\ &\leq A[g](x,\xi_1) + A[g](x,\xi_2) + \delta - \frac{n}{2}. \end{aligned}$$

By the arbitrariness of δ we have

$$A[g](x,\xi) < r^*(x,\xi_1) + r^*(x,\xi_2) - \frac{n}{2}.$$

Hence $A[g]$ is an admissible function.

The assumption (3.37) means

$$s_u(x,\xi) \geq A[g](x,\xi), \quad s_v(x,\xi) \geq A[g](x,\xi),$$

then Lemma 3.10 implies Lemma 3.11. □

Remark 3.4. When $f(x,u)$ is a polynomial of u with C^∞ coefficients, the inequality

$$s_f(x,\xi) \geq A[g](x,\xi) \tag{3.39}$$

also holds.

Lemma 3.12. *If $g(x,\xi)$ satisfies the assumptions in Lemma 3.11, $k > n/2$, then*

$$A[\min(g,k)] = \min(A[g],k). \tag{3.40}$$

Proof. Before proving Eq. (3.40) let us first prove that if $\sigma(x,\xi)$ is admissible function, then $\min(\sigma,k)$ is also admissible. Indeed, $\min(\sigma,k) > \frac{n}{2}$ is obvious. Take $\xi \in \widehat{\xi_1\xi_2}$, then for the case $\sigma(x,\xi_1) > k$, we have

$$\begin{aligned} \min(\sigma(x,\xi),k) &\leq k = \min(\sigma(x,\xi_1),k) \\ &\leq \min(\sigma(x,\xi_1),k) + \min(\sigma(x,\xi_2),k) - \frac{n}{2}. \end{aligned}$$

For the case $\sigma(x,\xi_2) > k$ the argument is similar. For the case $\sigma(x,\xi_1) \le k, \sigma(x,\xi_2) \le k$, we have

$$\min(\sigma(x,\xi),k) \le \sigma(x,\xi) \le \sigma(x,\xi_1) + \sigma(x,\xi_2) - \frac{n}{2}$$
$$\le \min(\sigma(x,\xi_1),k) + \min(\sigma(x,\xi_2),k) - \frac{n}{2}.$$

Moreover, $\min(\sigma(x,\xi),k)$ is a lower semi-continuous function, then it is admissible.

Choose any admissible function σ no more than g. Since $A[g]$ is the supremum of such functions, then

$$\sigma(x,\xi) \le A[g](x,\xi) \le g(x,\xi),$$

$$\min(\sigma(x,\xi),k) \le \min(g(x,\xi),k).$$

Because $\min(\sigma,k)$ is admissible, then $\min(\sigma,k) \le A[\min(g,k)]$. In view of the fact that $A[g]$ is the supremum of all admissible functions no more than g, then

$$\min(A[g],k) \le A[\min(g,k)].$$

The inequality in another direction is easier. Since $\min(g,k) \le g$, then $A[\min(g,k)] \le A[g]$. Similarly, $A[\min(g,k)] \le k$, then $A[\min(g,k)] \le \min(A[g],k)$.

Summing up, we obtain Eq. (3.40). □

Definition 3.4. Denote by v the solution of the linear wave equation with initial data Eq. (3.27). Let $r^{(k)}(x,\xi)$ $(k \ge 0)$ satisfies

$$r^{(0)}(x,\xi) = s_v(x,\xi), \tag{3.41}$$

$$r^{(k)}(x,\xi) = \min(s_v(x,\xi), \inf_{(x,\eta)\in\Gamma^-(x,\xi)} A[r^{(k-1)}](y,\eta) + 1) \quad (k > 0), \tag{3.42}$$

where $\Gamma^-(x,\xi)$ stands for the downward bicharacteristic strip issuing from (x,ξ). Then the limit

$$r(x,\xi) = \lim_{k\to\infty} r^{(k)}(x,\xi) \tag{3.43}$$

is called **regularity index**.

Remark 3.5. In the definition it is easy to see that $r^{(k)}(x,\xi)$ is a bounded and monotone decreasing function with respect to k, then the limit exists, so that the index of regularity is well defined.

Theorem 3.6. *Assume that $f(x,u)$ is a polynomial of u with C^∞ coefficients. $\phi_0 \in H^s, \phi_1 \in H^{s-1}$ $(s > n/2)$, then the Cauchy problem (3.26), (3.27) admits an H^s solution, the regularity function of it satisfies*

$$s_u(x,\xi) \geq r(x,\xi), \tag{3.44}$$

where $r(x,\xi)$ is the regularity index determined by the data (3.27).

Proof. Denote by v the solution of the homogeneous wave equation with initial data (3.27), by E the forward fundamental solution operator, i.e. Eg is the solution of the equation $\Box u = g$ satisfying initial conditions

$$u\big|_{x_n=0} = (\partial_{x_n} u)\big|_{x_n=0} = 0 \tag{3.45}$$

in $x_n > 0$. Then the solution of the problem (3.26), (3.27) can be written as

$$u = v + Ef(x,u). \tag{3.46}$$

By using Eq. (3.46) we can write u by an iteration form as

$$u_0 = u, \quad u_k = v + Ef(x, u_{k-1}) \text{ for } k > 0. \tag{3.47}$$

Obviously, when u is the solution of the problem (3.26), (3.27), then all u_k are the same as u. Then we can estimate the regularity of u via that of u_k. To simplify notations we write $f_k = f(x,u_k)$. Definition 3.3 implies $s_v(x,\xi) > r^{(0)}(x,\xi)$. From $u \in H^s$ we know

$$s_{u_0}(x,\xi) \geq s > n/2,$$

which implies $s_{f_0} > n/2$ and then $s_{Ef_0} > n/2+1$. From $u_1 = v + Ef_0$ we know

$$s_{u_1}(x,\xi) \geq \min(s_v(x,\xi), s_{Ef_0}(x,\xi)) \geq \min\left(r^{(0)}(x,\xi), \frac{n}{2}+1\right).$$

Generally, if we assume

$$s_{u_k}(x,\xi) \geq \min\left(r^{(k-1)}(x,\xi), \frac{n}{2}+k\right), \tag{3.48}$$

then Lemmas 3.10 and 3.11 imply

$$s_{f_k}(x,\xi) \geq A\left[\min\left(r^{(k-1)}(x,\xi), \frac{n}{2}+k\right)\right] = \min\left(A[r^{(k-1)}](x,\xi), \frac{n}{2}+k\right).$$

By using the theorem of singularity propagation for linear equations, we have

$$s_{Ef_k}(x,\xi) \geq \min\left(\inf_{(y,\eta)\in\Gamma^-(x,\xi)} A[r^{(k-1)}](y,\eta)+1, \frac{n}{2}+k+1\right).$$

By using $u_{k+1} = v + Ef(x, u_k)$,

$$\begin{aligned} s_{u_{k+1}}(x,\xi) &\geq \min(s_v(x,\xi), s_{Ef_k}(x,\xi)) \\ &\geq \min\left(r^{(0)}(x,\xi), \inf_{(y,\eta)\in\Gamma^-(x,\xi)} A[r^{(k-1)}](y,\eta)+1, \frac{n}{2}+k+1\right) \\ &= \min\left(r^{(k)}(x,\xi), \frac{n}{2}+k+1\right). \end{aligned}$$

Then Eq. (3.48) is valid for all k. Now if $r(x,\xi)$ is finite, by taking k sufficiently large,

$$s_u(x,\xi) = s_{u_{k+1}}(x,\xi) \geq r^{(k)}(x,\xi) \geq r(x,\xi). \tag{3.49}$$

Obviously, if $r(x,\xi)$ is $+\infty$, then $r^{(k)}(x,\xi)$ is also $+\infty$. Hence Eq. (3.44) is valid. □

Next we give an example, from which we can get more understanding on the singularity interaction.

Example 3.1. Let $\omega_i (i = 1, 2, 3)$ are three points in S^{n-2}, different from each other. On $x_n = 0$ the initial data (3.27) has singularity at $(\omega_i, 0)$. The regularity function $s_v(x,\xi)$ of the solution v to the linear wave equation with the initial condition (3.27) satisfies

$$s_v(x,\xi) \geq \begin{cases} s_0 > \frac{n}{2}, & (x,\xi) \in \Gamma_1 \cup \Gamma_2 \cup \Gamma_3, \\ +\infty, & \text{otherwise}, \end{cases} \tag{3.50}$$

where Γ_i is the bicharacteristic strip in $R^n_x \times R^n_\xi$, $\Gamma_i = \{(\omega_i(1 - x_n), x_n; \omega_i, 1)\}$. The question is how does the singularity of the solution to the nonlinear problem (3.26), (3.27).

In order to determine the singularity of the solution u to the nonlinear problem, we construct the regularity index. Denote by P_i the point $\left(\frac{\omega_i}{\sqrt{2}}, \frac{1}{\sqrt{2}}\right)$, let

$$r^{(0)}(x,\xi) = \begin{cases} s_0 > \frac{n}{2}, & \text{if } (x,\xi) \in \cup\Gamma_i; \\ \infty, & \text{otherwise.} \end{cases}$$

$$A[r^{(0)}](x,\xi) = \begin{cases} s_0, & (x,\xi) \in \bigcup \Gamma_i, \\ 2s_0 - \frac{n}{2}, & x = x_0, \xi \in \cup \overset{\frown}{P_i P_j}, \\ 3s_0 - n, & x = x_0, \xi \in \Delta P_1 P_2 P_3, \end{cases}$$

where $\Delta P_1P_2P_3$ is the spherical triangle enclosed by three arcs $\widehat{P_iP_j}$. When ξ locates in the arc $\widehat{P_iP_j}$ or $\Delta P_1P_2P_3$, (x,ξ) is not in the characteristic set of the wave equation, then these singularities will not propagate out. Therefore, we have

$$r^{(1)}(x,\xi) = \min(A[r^{(0)}](x,\xi)+1, r^{(0)}(x,\xi)),$$

$$A[r^{(1)}](x,\xi) = A[r^{(0)}](x,\xi)$$

$$\cdots\cdots$$

$$r(x,\xi) = \min(A[r^{(0)}](x,\xi)+1, r^{(0)}(x,\xi)),$$

Then we obtain the estimate of the regularity function of u:

$$s_u(x,\xi) \geq \begin{cases} s_0, & (x,\xi)\in\bigcup\Gamma_i; \\ 2s_0-\frac{n}{2}+1, & x=x_0, \xi\in\cup P_iP_j; \\ 3s_0-n+1, & x=x_0, \xi\in\Delta P_1P_2P_3, \\ +\infty, & \text{otherwise}. \end{cases} \tag{3.51}$$

The above calculation indicates that the singularity of u still stay on Γ_i with $i=1,2,3$. In this case, no regularity caused by interaction propagate out, even though three bicharacteristics bearing singularity meet together.

However, if we keep Γ_1 and Γ_2, and change Γ_3 to $\{(\omega_3(1-x_n), x_n; -\omega_3, -1)\}$, then the situation will be different. Because in this case the triangle $\Delta P_1P_2P_3$ on the unit sphere in the cotangent bundle based on x_0 may contain a direction ξ, which let (x_0,ξ) belong to the characteristic set of the operator $\Box$. Indeed, if we connect P_3 and any point in the arc P_1P_2 by great circle, then the great circle will intersect the characteristic set $\{(x_0; \frac{\omega}{\sqrt{2}}, \frac{1}{\sqrt{2}}); \omega\in S^{n-2}\}$. Denote the set of intersections by $\{\tilde\gamma\}$, then $A[r^{(0)}]$ is at least H^{3s_0-n} regular on $\{\tilde\gamma\}$. Therefore, denote

$$\tilde\Gamma = \left((1-x_n)\omega_0, x_n; \frac{\omega_0}{\sqrt{2}}, \frac{1}{\sqrt{2}}\right),$$

where $x_n>1, \omega_0\in S^{n-2}$ satisfying $(x_0; \frac{\omega}{\sqrt{2}}, \frac{1}{\sqrt{2}})\in\tilde\gamma$, we have

$$r^{(1)}(x,\xi) = \begin{cases} s_0, & (x,\xi)\in\bigcup\Gamma_i; \\ 2s_0-\dfrac{n}{2}+1, & x=x_0, \xi\in\cup P_iP_j; \\ 3s_0-n+1, & x=x_0, \xi\in\Delta P_1P_2P_3 \text{ or } (x,\xi)\in\{\tilde\Gamma\}; \\ +\infty, & \text{otherwise}. \end{cases}$$

$$r^{(2)}(x,\xi) = r^{(1)}(x,\xi).$$

$$\cdots\cdots$$

$$r(x,\xi) = r^{(1)}(x,\xi).$$

Therefore,

$$s_u(x,\xi) \geq \begin{cases} s_0, & (x,\xi) \in \bigcup \Gamma_i; \\ 2s_0 - \dfrac{n}{2} + 1, & x = x_0, \xi \in \cup P_i P_j; \\ 3s_0 - n + 1, & x = x_0, \xi \in \Delta P_1 P_2 P_3 \ \text{ or } (x,\xi) \in \{\tilde{\Gamma}\}; \\ +\infty, & \text{otherwise}. \end{cases} \tag{3.52}$$

Equation (3.52) means that the characteristic strips $\tilde{\Gamma}$ may possibly bear the singularities of u. The extra singularity caused by the interaction propagates out.

The above discussion indicates that for a given H^s solution u to a nonlinear problem, due to the double effect caused by linear propagation and nonlinear interaction the singularity of solution may diffuse to a bigger set than that for corresponding linear problem. We emphasize here that the above discussion in this chapter only gives the possibility of the appearance of singularities on corresponding bicharacteristics, but does not confirm that the singularity must appear there. To describe the set of singularity of solution precisely more careful analysis is required. In [7] a solution with the singularity as described in the above example is constructed.

For the production of extra singularities of solutions M.Beals proved the following theorem:

Theorem 3.7. *Assume that $s > (n+1)/2, n > 1$, then there is $\beta \in C_0^\infty(R^{n+1})$, and a compactly supported function $u_0 \in H^s(R^n), u_1 \in H^{s-1}(R^n)$, satisfying* sing supp $(u_0, u_1) = \{0\}$, *such that the solution $u \in H^s((0,1) \times R^n)$ to the Cauchy problem*

$$\begin{cases} (\partial_t^2 - \sum_{i=1}^{n} \partial_{x_i}^2)u = \beta u^3; \\ u|_{t=0} = u_0, \quad u_t|_{t=0} = u_1 \end{cases} \tag{3.53}$$

has its singular points with singularity of $H^{3s-n+2+\epsilon}$ full of the whole cone $\{(t,x); |x| < t, 0\}$.

It is well known that all bicharacteristics issuing from the origin for the wave operator locate on the characteristic cone with its vertex at the origin. Hence according to the rule of singularity propagation for the solutions to linear equations the singularity of the initial data at the origin may at most propagate to the surface of the characteristic cone. However, Theorem 3.7 indicates that for the solutions to the semilinear equations the $H^{3s-n+2+\epsilon}$ may spread to the whole solid cone. It means that for the H^s solutions to the semilinear wave equation, as well as to more general nonlinear equations, the $3s$ singularity propagation theorem is the best theorem one can expect. Therefore, to describe the propagation and interaction of weaker singularities of solutions some new concepts and methods should be introduced.

3.4 Propagation of conormal singularity

In this section a new concept called **conormal singularity** will be introduced. Conormal singularity can be used to describe the solutions having singularity on some surface, where the singularity of solutions is comparatively stronger on the normal direction, and is much weaker on the tangential direction. A distribution having conormal singularity is called **conormal distribution**. In the sequel we will give its precise definition. As we will see that such a concept is suitable to describe the propagation of the classical wave fronts in the study of wave motion.

Definition 3.5. Assume that $\mathscr{S} : (S_1, \cdots, S_n)$ is a set of C^∞ submanifolds in $\Omega \subset R^n$, $\mathscr{V}(\mathscr{S})$ is a set of $C^\infty(\Omega, T(\Omega))$ vector fields, tangential to each S_i in $\mathscr{S}$. If $\mathscr{V}(\mathscr{S})$ forms a Lie algebra, i.e. $V_1, V_2 \in \mathscr{V}$ implies $[V_1, V_2] \in \mathscr{V}$, then $\mathscr{V}(\mathscr{S})$ is called **complete**.

The definition of completeness of $\mathscr{V}(\mathscr{S})$ is independent of the choice of coordinates. The fact is obvious from the meaning of its definition. Next we verify it by calculation for the case when $\mathscr{S}$ only contains one surface S_1. Assume that the equation of S_1 is $f(x) = 0$. The vector fields $V_1 = \sum a_i \frac{\partial}{\partial x_i}, V_2 = \sum b_i \frac{\partial}{\partial x_i}$ belong to $\mathscr{V}(\mathscr{S})$. Then $\sum a_i \frac{\partial f}{\partial x_i} = \sum b_i \frac{\partial f}{\partial x_i} = 0$, while $[V_1, V_2] \in \mathscr{V}(\mathscr{S})$ means

$$\sum_{i,j} \left(a_i \frac{\partial b_j}{\partial x_i} - b_i \frac{\partial a_j}{\partial x_i} \right) \frac{\partial f}{\partial x_j} = 0. \tag{3.54}$$

Under the coordinate transformation $x = x(y)$ the equation of S_1 becomes $g(y) = f(x(y)) = 0$, the vector fields V_1, V_2 become $\sum \tilde{a}_k \frac{\partial}{\partial y_k}, \sum \tilde{b}_k \frac{\partial}{\partial y_k}$, where $\tilde{a}_k = \sum a_i \frac{\partial y_k}{\partial x_i}, \tilde{b}_k = \sum b_i \frac{\partial y_k}{\partial x_i}$, then

$$\sum_{k,\ell} \left(\tilde{a}_k \frac{\partial \tilde{b}_\ell}{\partial y_k} - \tilde{b}_k \frac{\partial \tilde{a}_\ell}{\partial y_k} \right) \frac{\partial g}{\partial y_\ell} = \sum_{i,h,\ell} \left(a_i \frac{\partial \tilde{b}_\ell}{\partial x_i} - b_i \frac{\partial \tilde{a}_\ell}{\partial x_i} \right) \frac{\partial f}{\partial x_h} \frac{\partial x_h}{\partial y_\ell}$$

$$= \sum_{i,j,h,\ell} \left(a_i \frac{\partial b_j}{\partial x_i} \frac{\partial y_\ell}{\partial x_j} - b_i \frac{\partial a_j}{\partial x_i} \frac{\partial y_\ell}{\partial x_j} \right) \frac{\partial f}{\partial x_h} \frac{\partial x_h}{\partial y_\ell}$$

$$= \sum_{i,j,h,\ell} \left(a_i \frac{\partial b_j}{\partial x_i} - b_i \frac{\partial a_j}{\partial x_i} \right) \frac{\partial f}{\partial x_h} \delta_{hj} = 0.$$

Hence in the new coordinates $[V_1, V_2]$ is still tangential to S_1.

Definition 3.6. If $\mathscr{V}(\mathscr{S})$ is a complete tangential vector field, the function $u \in H^s_{loc}(\Omega)$, and for any $i \leq k$ and $V_1, \cdots, V_i \in \mathscr{V}(\mathscr{S})$, we have

$$V_1 \cdots V_i u \in H^s_{loc}(\Omega), \tag{3.55}$$

then u is called **conormal distribution of order** k with respect to $\mathscr{S}$, and is denoted by $u \in H^{s,k}(\Omega, \mathscr{S})$, which is called **space of conormal distributions**.

As we mentioned above, singularity is a local property for given functions, then we often omit the subscript "loc" later.

In the set $\mathscr{V}(\mathscr{S})$ of conormal distributions we can choose a base B : $\{\tilde{V}_1, \cdots, \tilde{V}_N\}$, such that for any vector field $V \in \mathscr{V}$, one can find finite vector fields $\tilde{V}_{i_1}, \cdots, \tilde{V}_{i_\ell}$ in B such that V can be written as the linear combination of them:

$$V = \sum_{s=1}^{\ell} a_s \tilde{V}_{i_s},$$

where all a_s are C^∞ functions on Ω. Then the completeness of $\mathscr{V}$ can be described as : for any $V_i, V_j \in B$, there are C^∞ coefficients C^k_{ij}, such that

$$[V_i, V_j] = \sum_k C^k_{ij} V_k. \tag{3.56}$$

Obviously, it is enough that in Definition 3.6 the vector fields $V_1, \cdots, V_i$ can be chosen in a given basis. Moreover, we can also define the space $H^{s,k}(\Omega, \mathscr{S})$ by the norm

$$\|u\|_{H^{s,k}(\Omega,\mathscr{S})} = (\sum \tilde{V}_{i_1} \cdots \tilde{V}_{i_\ell} u)^{1/2}, \tag{3.57}$$

where $\ell \le k$, $\tilde{V}_{i_1}, \cdots, \tilde{V}_{i_\ell}$ are all possible different choices in the basis. It is evident that the norm is equivalent, if we choose different base for a given space $H^{s,k}(\Omega, \mathscr{S})$.

Any function $u \in H^{s,k}(\Omega, \mathscr{S})$ is H^{s+k} regular outside $\mathscr{S}$. Besides, if $x \in \mathscr{S}$ and $\mathscr{S}$ is a smooth hypersurface S_0 in a neighborhood of x, then u is H^s microlocal regular in the normal direction of S_0 and is H^{s+k} microlocal regular in the tangential direction of S_0. Since s and k are two independent numbers, then the concept of conormal distribution can be applied to study the singularity propagation for those distributions, which have weak singularity in some specific direction.

Next let us give some examples of conormal distributions.

Example 3.2. If $\mathscr{S}$ is a single hypersurface $S_1 : x_1 = 0$, then the base of $\mathscr{V}(\mathscr{S})$ can be chosen as

$$x_1 D_{x_1}, D_{x_2}, \cdots, D_{x_n}.$$

Direct calculation indicates that $\mathscr{V}(\mathscr{S})$ is complete. According to the definition, the space $H^{s,k}(\Omega, S_1)$ is composed of all functions satisfying

$$(x_1 D_{x_1})^{\alpha_1} D_{x_2}^{\alpha_2} \cdots D_{x_n}^{\alpha_n} u \in H^s(\Omega), \quad \alpha_1 + \cdots + \alpha_n \le k.$$

Example 3.3. If $\mathscr{S}$ is composed of two hypersurfaces $S_1 : x_1 = 0, S_2 : x_2 = 0$ and their intersection $x_1 = x_2 = 0$, then the base of $\mathscr{V}(\mathscr{S})$ can be chosen as

$$x_1 D_{x_1}, x_2 D_{x_2}, D_{x_3}, \cdots, D_{x_n}.$$

Direct calculation also indicates that $\mathscr{V}(\mathscr{S})$ is complete. According to the definition, the space $H^{s,k}(\Omega, S)$ is composed of all functions satisfying

$$(x_1 D_{x_1})^{\alpha_1} (x_2 D_{x_2})^{\alpha_2} \cdots D_{x_n}^{\alpha_n} u \in H^s(\Omega), \quad \alpha_1 + \cdots + \alpha_n \le k.$$

Like the discussion in the above sections of this chapter the closeness of the set of functions with specific singularity under nonlinear composition is crucial in the study of singularity propagation for the solutions to nonlinear equations. Therefore, we are going to prove the following lemma now:

Lemma 3.13. *Let $\mathscr{V}(\mathscr{S})$ is a set of vector fields mentioned above. $s > n/2$, k is an integer, $u, v \in H^{s,k}$, $f(x, u)$ is a C^∞ function of its arguments, then $uv \in H^{s,k}$ and $f(x, u) \in H^{s,k}$.*

Proof. Assume $k = 1$, then by acting an operator V in $\mathscr{V}$ on uv, we have

$$V(uv) = uVv + vVu. \tag{3.58}$$

By using Schauder lemma the right hand side belongs to H^s. In view of the fact that V is an arbitrary element in $\mathscr{V}$, then $uv \in H^{s,1}$. For the case $k > 1$ we can use Leibnitz formula to establish $uv \in H^{s,k}$.

Acting V on $f(x, u)$ gives

$$Vf(x,u) = f_x(x,u) + f_u(x,u)Vu. \tag{3.59}$$

Obviously, $Vu \in H^s$. Moreover, $f_x(x,u), f_u(x,u) \in H^s$ due to $u \in H^s$. Therefore, the right hand side of Eq. (3.59) in is H^s, so that $f(x,u) \in H^{s,1}$. Now if for any $k_1 < k$ any nonlinear composition belongs to $H^{s,1}$, then the right hand side of Eq. (3.59) also belongs to H^{s,k_1}. This leads to $f(x,u) \in H^{s,k_1+1}$. Hence $f(x,u) \in H^{s,k}$ by induction. □

Next we consider the propagation of conormal singularities of solutions to semilinear equation

$$Pu = F(x, u, \cdots, D^{m-1}u) \tag{3.60}$$

in Ω. Here P is a linear differential operator with C^∞ coefficients of order m, F is a C^∞ function of its arguments. Like the previous section we assume that $\Omega_+ = \Omega \cap \{x > 0\}$ is included in the determinacy domain of $\Omega_- = \Omega \cap \{x < 0\}$, and all eigenvalues of the principal symbol $p_m(x, \xi)$ of P are real and simple. Then we have

Theorem 3.8. *If S is a characteristic surface of the operator P in the domain Ω, u is $H^s(\Omega)$ solution of Eq. (3.60), $s > n/2 + m - 1$, then $u \in H^{s,k}(\Omega_-, S)$ implies $u \in H^{s,k}(\Omega_+, S)$.*

Proof. By a suitable transformation we may assume that the equation of S is $x_1 = 0$. Then the operator P can be written as

$$P = x_1 a_1(x) D_{x_1}^m + \sum_{i=2}^{n} A_j(x, D_x) D_{x_j} + A_0(x, D_x), \tag{3.61}$$

where $a_1(x) \in C^\infty$, $A_j(j = 0, 2, \cdots, n)$ are operators of order $m - 1$. For any base vector $V_1 = x_1 D_{x_1}$, $V_j = D_{x_j} (j \geq 2)$, we can write

$$[V_i, P] = B_{i0} + \sum_{j=1}^{n} B_{ij} V_j + B_i P, \tag{3.62}$$

where B_{ij} are differential operators with C^∞ coefficients of order $m - 1$, B_i are C^∞ functions. Equation (3.62) means that the commutator of V_i and P can be expressed as a combination of V_i and P. It is also called **commutator relation**.

Regarding I as V_0, $[V_0, P]$ can also be written as the form of Eq. (3.62), where corresponding B_0, B_{0j} are all zero. Hence, by acting the operator V_i with $0 \le i \le n$ on Eq. (3.60) we have

$$PV_i u + \sum_{j=0}^{n} B_{ij} V_j u = V_i F(x, u, \cdots .D^{m-1}u). \tag{3.63}$$

Let $U = {}^t(u, V_1 u, \cdots, V_n u)$, one can write the right-hand side of Eq. (3.63) as $F_i(x, U, \cdots, D^{m-1}U)$, where F_i are C^∞ functions of their arguments. Equation (3.63) can also be written as the matrix form

$$PU + BU = F(x, U, \cdots, D^{m-1}U), \tag{3.64}$$

where the left-hand side is a matrix of operators of m-th order. The principal symbol of this matrix is a diagonal matrix with $p_m(x, \xi)$ on its diagonal. According to the assumptions of the theorem $U \in H^s(\Omega_-)$. Then $U \in H^s(\Omega)$ according to the remark of Theorem 3.4, i.e. $u \in H^{s,1}(\Omega, S)$. Hence the conclusion of the theorem is valid for the case $k = 1$.

In order to discuss the case $k > 1$, we need to have a more general commutator relation, i.e.

$$[V^I, P] = \sum_{|J| \le |I|} B_{I,J} V^J + \sum_{|K| \le |I|-1} C_{I,K} V^K P, \tag{3.65}$$

where $V^I = V_{i_1} \cdots V_{i_\ell}, |I| = \ell$, $B_{I,J}$ are partial differential operators of order $m - 1$, $C_{I,K}$ are C^∞ functions. This commutator relation can be easily verified by induction. Indeed, if Eq. (3.65) is valid for the multi-index I satisfying $|I| = k$, then for I' satisfying $|I'| = k + 1$ we have

$$\begin{aligned}
[V^{I'}, P] &= [V_i V^I, P] = V_i [V^I, P] + [V_i, P] V^I \\
&= V_i \left(\sum_{|J| \le k} B_{I,J} V^J + \sum_{|K| \le |I|-1} C_{I,K} V^K P \right) \\
&\quad + \left(B_{i0} + \sum_{1 \le j \le n} B_{ij} V_j + B_i P \right) V^I \\
&= \sum_{|J| \le k+1} B'_{I,J} V^J + \sum_{|J| \le k} C'_{I,J} V^J P - B_i [V^I, P] + B_i V^I P.
\end{aligned}$$

Substituting Eq. (3.65) with $|I| = k$ into the right hand side we obtain such an equality with $|I| = k + 1$. Hence the general commutator relation Eq. (3.65) holds for any multi-index I by induction.

Having the general commutator relation (3.65), and letting U_k be the column vector ${}^t(u,\cdots,V^I u,\cdots)$, where all elements with the form $V_{i_1}\cdots V_{i_\ell}(\ell<k)$ are included, one can obtain the system

$$P_k U_k + B_k U_k = F_k(x, D^\beta U_k) \tag{3.66}$$

by acting the operator V^I on Eq. (3.60). Notice that the operator matrix in the left hand side of Eq. (3.66) has its symbol being diagonal matrix with same element $p_m(x,\xi)$, and in the right hand side F_k is C^∞ function of its arguments and only depends on the derivatives of U_k with order no more than $m-1$, then like the discussion in the case $k=1$ one can use the assumption $U_k \in H^s(\Omega_-)$ and Theorem 3.4 with its remark to obtain $U_s \in H^4(\Omega)$, i.e. $u \in H^{s,k}(\Omega, S)$. □

From the process of the proof we know that once the vector field $\mathscr{V}(S)$ is established, the key point to prove the rule of propagation of conormal singularities is the validity of the commutator relation. Therefore, the method in Theorem 3.8 is also called **commutator method**. Certainly, the completeness of $\mathscr{V}(S)$ and the closeness of the space of conormal distributions under nonlinear composition are necessary.

Next we discuss the case when two surfaces carrying cornomal singularities intersect. Let the domain Ω and the equation be the same as above, the characteristic surfaces S_1, S_2 of P intersect at Γ in Ω_+ transversally. Moreover, there are not other characteristic surfaces through Γ. Denoting by S the set $\{S_1, S_2, \Gamma\}$, by $\mathscr{V}(S)$ the set of all tangential vector fields with respect to S, then the following conclusion.

Theorem 3.9. *Assume that u is an $H^s(\Omega)$ solution of Eq. (3.60), $s > n/2+m-1$, the restriction of u on Ω_- belongs to $H^{s,k}(\Omega_-, S_1\cup S_2)$, then u belongs to $H^{s,k}(\Omega, S)$ in whole Ω.*

Proof. The proof can be proceeded like that for Theorem 3.8. Next we only verify some crucial points. As indicated in Example 2, $\mathscr{V}(S)$ is complete, then the space $H^{s,k}(\Omega,S)$ is well-defined and is closed under nonlinear composition. To prove the validity of the commutation relation we assume that the equations of S_1, S_2 are $x_1=0, x_2=0$ respectively, because otherwise we can arrive at the case by using a coordinate transformation. Since these are characteristics of the operator P, then P has the form

$$P = KD_{x_1}D_{x_2} + \sum_{j=1}^{n} A_j V_j + A_0, \tag{3.67}$$

where $V_1 = x_1 D_{x_1}, V_2 = x_2 D_{x_2}, V_j = D_{x_j}$ $(j \geq 3)$, the order of K is $m-2$ and the order of A_j $(1 \leq j \leq n)$ is $m-1$. Since the operator P does not have any other characteristic surface through Γ, then the symbol $k(x,\xi) \neq 0$ of K on $x_1 = x_2 = \xi_3 = \cdots = \xi_n = 0$. Denote by H the psuedodifferential operator of order $2-n$ with symbol $h(x,\xi) = (k(x,\xi) + \sum_{i=3}^{n} |\xi_i|^{m-2})^{-1}$, we have

$$HK = I + \sum_{j=1}^{n} R_j V_j + R_0, \tag{3.68}$$

where R_j are of order -1. Hence

$$D_{x_1} D_{x_2} = D_{x_1} D_{x_2} \left(HK - \sum_{j=1}^{n} R_j V_j + R_0 \right) = HP + \sum_{j=1}^{n} B_j V_j + B_0,$$

where $B_j (0 \leq j \leq n)$ are of first order. Therefore, according to the expression (3.67) of P the commutator $[P, V_j]$ can be written as

$$\begin{aligned} [P, V_j] &= \sum C_j V_j + C_0 + G D_{x_1} D_{x_2} \\ &= \sum C_j V_j + C_0 + GHP + \sum E_j V_j + E_0, \end{aligned}$$

where C_j, E_j are of order $m-1$, G is of order $m-2$.

Acting $V_1, \cdots, V_n$ on Eq. (3.60) and denoting $U_1 = {}^t(u, V_1 u, \cdots, V_n u)$, we obtain the system

$$P_1 U_1 + B_1 U_1 = L F_1(x, U_1, \cdots, D^{m-1} U_1), \tag{3.69}$$

where B_1 is a matrix of pseudodifferential operators of order $m-1$, L is a pseudodifferential operator of order. Comparing Eq. (3.69) with Eq. (3.64), the principal part is the same, and the new terms in the right hand side is a map from H^s to H^s, then we can use the same method to prove that $U_1 \in H^s(\Omega_-)$ implies $U_1 \in H^s(\Omega)$, so that $u \in H^{s,1}(\Omega, S)$.

Finally, the conclusion $u \in H^{s,k}(\Omega.S)$ can be derived by induction. □

Theorem 3.9 indicates that when two characteristic surface carrying conormal singularities intersect, the singularities propagate along the characteristics and the intersection will not produce any new singularities, provided no any other characteristic surface issues through the intersection. However, if there are other characteristic surfaces issuing through the intersection, then the extra singularity occurs. This is the subject to be discussed in the next section.

3.5 Interaction of conormal singularities

3.5.1 *Extension of the concept of conormal singularities*

In the previous section we see that in the study of the propagation of conormal singularities the commutator relation of a vector field $\mathscr{V}(S)$ with respect to a partial differential operator P plays a key role. However, if there are more than two characteristic surfaces intersecting at a same submanifold, such a commutator relation may not be valid.

Let us look an example. Consider the equation

$$Pu = f(x, u, Du, D^2u),$$

where $P = D_{x_1}D_{x_2}(D_{x_1} + D_{x_2})$. Obviously, P is hyperbolic with respect to the direction $(1, 1)$. If the solution u has conormal singularities on S_1 : $x_1 = 0$ and S_2 : $x_2 = 0$ when $x_1+x_2 < 0$, we hope to know the singularities of u for $x_1 + x_2 > 0$.

Let S_3 be the submanifold $x_1 = x_2$, Γ be the intersection $S_1 \cap S_2$. Denote by S the set of submanifolds S_1, S_2, S_3 and their intersection Γ. Let $\mathscr{V}(S)$ be the vector fields tangential to the set S, then the basis of $\mathscr{V}(S)$ can be found in the following way. If a vector field V is tangential to $S_!, S_2, \Gamma$, then it can be written as

$$a_1x_1D_{x_1} + a_2x_2D_{x_2} + \sum_{i=3}^{n} a_iD_{x_i}.$$

If V is also tangential to S_3, its symbol must vanish on the conormal bundle of S_3. This means that the symbol of V must vanish, provided $\xi_1 = -\xi_2, \xi_3 = \cdots = \xi_n = 0, x_1 = x_2$. Then there is a C^∞ function b satisfying

$$a_1x_1 - a_2x_2 = b(x_1 - x_2),$$

which implies

$$(a_1 - b)x_1 = (a_2 - b)x_2.$$

Since all coefficients are C^∞ functions, then there is a C^∞ function c, such that

$$a_1 - b = cx_2, \quad a_2 - b = cx_1.$$

It means that V can be written as

$$b(x_1D_{x_1} + x_2D_{x_2}) + cx_1x_2(D_{x_1} + D_{x_2}) + \sum_{i=3}^{n} a_iD_{x_i}. \tag{3.70}$$

Conversely, any vector field with the form (3.70) must be tangential to any sudmanifold in S, so that it belongs to $\mathscr{V}(S)$. Therefore, the basis of $\mathscr{V}(S)$ can be chosen as

$$V_1 = x_1 D_{x_1} + x_2 D_{x_2}, \quad V_2 = x_1 x_2 (D_{x_1} + D_{x_2}), \quad V_i = D_{x_i} \ (i \geq 3). \quad (3.71)$$

Now consider the commutator of P and V_2. By using the notation of Poisson bracket

$$\{f(x,\xi), g(x,\xi)\} = \sum_{j=1}^{n} \left(\frac{\partial f}{\partial \xi_j} \frac{\partial g}{\partial x_j} - \frac{\partial f}{\partial x_j} \frac{\partial g}{\partial \xi_j} \right),$$

we have

$$\begin{aligned} Sym\ [P, V_2] &= \{\xi_1\xi_2(\xi_1+\xi_2), x_1x_2(\xi_1+\xi_2)\} \\ &= 2(x_1+x_2)\xi_1\xi_2(\xi_1+\xi_2) + (\xi_1+\xi_2)(x_1\xi_1^2 + x_2\xi_2^2) \\ &= 2(x_1+x_2)P + (\xi_1+\xi_2)(x_1\xi_1^2+x_2\xi_2^2). \end{aligned}$$

Obviously, if the commutator relation holds, then $(\xi_1+\xi_2)(x_1\xi_1^2+x_2\xi_2^2)$ can be expressed as a linear combination of the symbol of P, V_1, V_2 with C^∞ coefficients. Since $Sym\ P$, $Sym\ V_2$, $Sym\ [P, V_2]$ contain a common factor $\xi_1+\xi_2$, but $Sym\ V_1$ does not, then $[P, V_2]$ should be expressed by the linear combination of P and V_2. It means that $x_1\xi_1^2+x_2\xi_2^2$ is a linear combination of x_1x_2 and $\xi_1\xi_2$. However, this is impossible by a simple checking.

Therefore, to analyze the conormal singularities of solutions to partial differential equations one must understand the commutator relation in a generalized sense. That is, in the above commutator relation (3.62) the coefficients may not be a function or a differential operator. The pseudodifferential operator coefficients should also be allowed. Meanwhile, it is not necessary to let Eq. (3.62) hold in local sense. Instead, It holds in microlocal sense is enough. Hence we have to explain the concept of conormal distributions again, including the meaning of closeness of space of conormal distributions under nonlinear composition.

Let us start from the definition of the space of conormal distributions. Denote by OpS^k the class of pseudodifferential operators of order k. For OpS^1 operators $M_1, \cdots, M_N$ one can construct a set of operators $\mathscr{M}$, which is a linear combination with 1 and all $\{M_j\}$ as its basis and OpS^0 as coefficients. That is

$$M \in \mathscr{M} \Leftrightarrow M \in OpS^1, \ M = \sum_{j=1}^{N} A_j M_j + A_0 \ \text{ with } A_j \in OpS^0. \quad (3.72)$$

The set $\mathscr{M}$ is defined in Eq. (3.72) is called **pseudodifferential operators modulo.** Obviously, the principal symbol of a pseudodifferential operator determines that whether the operator belongs to $\mathscr{M}$. For a given operator L, if there is another operator $M \in \mathscr{M}$ such that the principal symbol of $M - L$ in a neighborhood of x_0 is zero (in a microlocal neighborhood of (x_0, ξ_0) resp.), we say L belongs to $\mathscr{M}$ locally at x_0 (microlocally at (x_0, ξ_0) resp.). Similarly, for given operators $\{L_1, \cdots, L_N\}$, if for any operator $M \in \mathscr{M}$, there are operators $A_j \in OpS^0$ $(1 \leq j \leq N)$, such that the principal symbol of $M - \sum A_j L_j$ in a neighborhood of x_0 is zero (in a microlocal neighborhood of (x_0, ξ_0) resp.), we say $\{L_1, \cdots, L_N\}$ produces $\mathscr{M}$ locally at x_0 (microlocally at (x_0, ξ_0) resp.).

Equation (3.72) is similar to the linear expression introduced in the last section to construct $\mathscr{V}$ by using $V_1, \cdots, V_N$, but the difference is that all A_j in Eq. (3.72) can be chosen as pseudodifferential operators of order zero, while the corresponding coefficients in the last section are restricted to C^∞ functions. In accordance, we introduce the second definition of conormal distribution as follows.

Definition 3.7. Let $\mathscr{M}$ be the pseudodifferential operators modulo defined by Eq. (3.72). If it forms a Lie algebra with respect to the commutator operation of pseudodifferential operators, then the **space of conormal distributions** $H^{s,k}(\Omega, \mathscr{M})$ is the set

$$u \in H^{s,k}(\Omega, \mathscr{M}) \Leftrightarrow u \in H^s, \ M_{i_1} \cdot M_{i_\ell} u \in H^s, \tag{3.73}$$

where $\ell \leq k, M_{i_j} \in \mathscr{M}$. Any element is $H^{s,k}(\Omega, \mathscr{M})$ is called **conormal distribution**.

Let us make a comparison of Definition 3.4 and Definition 3.5. For a given set $\mathscr{S}$ of submanifolds $(S_1, \cdots, S_N)$, the corresponding set $\mathscr{M}$ in Definition 3.5 can be chosen as the set of pseudodifferential operators, whose principal symbol vanishes on conormal bundles $\{N^*(S), S \in \mathscr{S}\}$. Obviously, any vector field in $\mathscr{V}(\mathscr{S})$ vanishes on conormal bundles $N^*S_1, \cdots, N^*S_N$, then $\mathscr{V} \subset \mathscr{M}$. It leads to $H^{s,k}(\Omega, \mathscr{M}) \subseteq H^{s,k}(\Omega, \mathscr{V})$. However, the inverse conclusion is true for the case when $\mathscr{S}$ is a single surface or two surfaces intersecting transversally, but it may not be true in some other cases.

If $\mathscr{S}$ is a single surface S, by a suitable coordinates transformation S can be transformed to $x_1 = 0$. Then the basis of $\mathscr{V}(S)$ has been shown in Example 1 of the last section. On the other hand the principal symbol $p(x, \xi)$ of any element in the modulo $\mathscr{M}$ satisfies

$$p(x, \xi) = 0, \quad \text{if } \ x_1 = 0, \ \xi_2 = \cdots = \xi_n = 0. \tag{3.74}$$

Therefore, $p(x,\xi)$ can be written as

$$a_1(x,\xi)x_1 + \sum_{i=2}^{n} a_i(x,\xi)\xi_i, \tag{3.75}$$

where a_1 is a homogeneous symbol of degree 1, a_i $(i \geq 2)$ are homogeneous symbols of degree 0. Writing a_1 as $\sum_{i=1}^{n} b_i\xi_i$ and incorporating the terms with $i \geq 2$ into the summation in the right hand side of Eq. (3.75), we obtain

$$p(x,\xi) = b_1(x,\xi)x_1\xi_1 + \sum_{i=2}^{n} a_i'(x,\xi)\xi_i. \tag{3.76}$$

It means that any element in $\mathscr{M}$ can be expressed by using the basis of $\mathscr{V}$ according to the requirement of Eq. (3.70). Hence

$$H^{s,k}(\Omega,\mathscr{M}) = H^{s,k}(\Omega,\mathscr{V}).$$

Consider the case $\mathscr{S} = S_1 \cup S_2 \cup (S_1 \cap S_2)$, where S_1 and S_2 intersect transversally. By using a suitable transformation the surface S_1 and S_2 can be reduced to $x_1 = 0$ and $x_2 = 0$ respectively. The basis of $\mathscr{V}(\mathscr{S})$ has been shown in Example 2 of the last section. On the other hand the principal symbol $p(x,\xi)$ of any element in $\mathscr{M}$ should satisfies

$$p(x,\xi) = 0, \quad \text{if} \quad \begin{cases} x_1 = 0, \xi_2 = \cdots = \xi_n = 0; \\ x_2 = 0, \xi_1 = \xi_3 = \cdots = \xi_n = 0; \\ x_1 = x_2 = 0, \xi_3 = \cdots = \xi_n = 0. \end{cases} \tag{3.77}$$

From the first condition of (3.77) $p(x,\xi)$ can be written as

$$b_1(x,\xi)x_1\xi_1 + \sum_{i=2}^{n} a_i'(x,\xi)\xi_i.$$

By applying the second condition of (3.77) the symbol $p(x,\xi)$ can be written as

$$b_1'(x,\xi)x_1\xi_1 + b_2'(x,\xi)x_2\xi_2 + c(x)\xi_1\xi_2 + \sum_{i=3}^{n} a_i''(x,\xi)\xi_i.$$

Furthermore, the third condition in (3.77) implies that $c(x)$ can be written as $c_1x_1 + c_2x_2$. It means that the term $c(x)\xi_1\xi_2$ also has the form as shown in the first two terms, so that it can be incorporated into the first two terms. Therefore, $\mathscr{M}$ can be expressed by using the basis of $\mathscr{V}$ according to the requirement of Eq. (3.72). Hence

$$H^{s,k}(\Omega,\mathscr{M}) = H^{s,k}(\Omega,\mathscr{V}).$$

However, for more general set of submanifolds, the space $H^{s,k}\Omega, \mathscr{M})$ and the space $H^{s,k}(\Omega, \mathscr{V})$ can be different. For instance, for $\mathscr{S}$ introduced in the beginning of this section the basis of tangential vector field $\mathscr{V}(\mathscr{S})$ is shown in (3.70). However, the symbol $a(x,\xi) = x_1|\xi|^{-1}\xi_1(\xi_1+\xi_2)$ is the annihilator of $T^*(S_i)$ and $T^*(\Gamma)$, but it cannot be expressed as a linear combination of symbols of operators in (3.69). Therefore, in this case $\mathscr{V}$ is strictly included in $\mathscr{M}$, so that $H^{s,k}(\Omega, \mathscr{M})$ is strictly included in $H^{s,k}(\Omega, \mathscr{V})$.

Next we proceed our discussion in the whole R^n, i.e. Ω is taken as R^n. For the notational simplicity, $H^{s,k}(R^n, \mathscr{M})$ is denoted by $H^{s,k}$. We notice that such a space of conormal distribution is not closed under non-linear composition, even is not closed under multiplication. Indeed, in two dimensional case consider the modulo $\mathscr{M}$ produced by pseudodifferential operator $(I-\Delta)^{-1/2}D_{x_1}D_{x_2}$ and construct the space $H^{s,\infty}$ by using this modulo. Obviously, by suitably taking the positive number α, we can let the functions $|x_1|^\alpha, |x_2|^\alpha \in H^{s,\infty}$, but $|x_1|^\alpha \cdot |x_2|^\alpha \notin H^{s,\infty}$. Therefore, in order to let the space of conormal distributions is closed under nonlinear composition we have to add some restrictions to $\mathscr{M}$.

Definition 3.8. Assume that $\mathscr{M}$ is a modulo of pseudodifferential operators. If for any $x_0 \in R^n$, and any $\xi^{(1)}, \xi^{(2)}, \eta \in R^n \setminus 0$, there exist vector fields $Z_1, \cdots, Z_q$ in a neighborhood of x_0, such that

(1) Z_j $(1 \le j \le q)$ belongs to $\mathscr{M}$ microlocally in neighborhoods of $(x^{(0)}, \xi_1)$ and $(x^{(0)}, \xi_2)$;

(2) For any $M \in \mathscr{M}$, there exists $A_j \in OpS^0$ $(1 \le j \le q)$, such that $Sym(M - \sum A_j Z_j) = 0$ in a neighborhood of (x_0, η).

Then one says that $\mathscr{M}$ satisfies **three points condition**.

Theorem 3.10. *If the modulo $\mathscr{M}$ satisfies the three points condition, then $H^{s,k}(\mathscr{M})$ with $s > n/2$ forms an algebra.*

Proof. Since $\mathscr{M}$ satisfies the three points conditions, then for any $(\xi^{(1)}, \xi_2^{(2)}, \eta) \in S^{n-1} \times S^{n-1} \times S^{n-1}$ we can find a system $\{Z_i\}$, such that the conditions (1) and (2) in Definition 3.6 hold in a neighborhood $O_{j_1} \times O_{j_2} \times O_{j_3}$ of $(\xi^{(1)}, \xi_2^{(2)}, \eta)$. Due to the compactness of S^{n-1} there is a system of finite open conical neighborhoods, which form an open covering to $S^{n-1} \times S^{n-1} \times S^{n-1}$. According to this open covering we can make a partition of unit for pseudodifferential operators $1 = \sum E_j(x, D_x)$, where $E_j \in OpS^0$, and the support of E_j is included in the open neighborhood.

Therefore, for any u, v

$$M(uv) = \sum_{j_1} \sum_{j_2} \sum_{j_3} E_{j_3} M(E_{j_1} u \cdot E_{j_2} v), \tag{3.78}$$

where the typical term can be written as

$$DM(Bu \cdot Cv) = \sum D_j(Z_j Bu \cdot Cv) + \sum D_j(Bu \cdot Z_j Cv).$$

Since Z_j belongs to $\mathscr{M}$ microlocally on the support of $Sym(B)$ or $Sym(C)$, then $Z_j B, Z_j C \in \mathscr{M}$. Hence by using the basis of $\mathscr{M}$ the term $M(uv)$ can be expressed as a function $\mathscr{L}(u, v, M_i u, M_i v)$, which is obtained by acting pseudodifferential operators of order 0 on $u, v, M_i u, M_i v$ and making a nonlinear composition. Since $u, v, M_i u, M_i v \in H^s$, any pseudodifferential operator of order 0 is a continuous map from H^s to H^s, and H^s is closed with respect to nonlinear composition, then $M(uv) \in H^s$, which implies $uv \in H^{s,1}$. It means that $H^{s,1}$ forms an algebra. Furthermore, the fact that $H^{s,k}$ forms an algebra can be proved by induction. □

Let us give some examples to explain the three points conditions.

Example 3.4. Let $\mathscr{S}$ be composed by Γ and the surfaces $S_1, \cdots, S_N$ passing through Γ. $\mathscr{M}$ is composed by all pseudodifferential operators of order 1, whose symbol annihilate $N^*(\Gamma)$ and all $N^*(S_j)$, then $\mathscr{M}$ satisfies three points condition.

Indeed, for any three points $(x_0, \xi^{(1)}), (x_0, \xi^{(2)}), (x_0, \eta)$, $\mathscr{M}$ satisfies the three points condition is trivial for $x_0 \notin \Gamma$. If $x_0 \in \Gamma$, then there are at least $N-2$ surfaces among $S_1, \cdots, S_N$ not perpendicular to $\xi^{(1)}$ and $\xi^{(2)}$. Without loss of generality we assume these $N-2$ surfaces are $S_3, \cdots, S_N$. Denote $S' = S_1 \cup S_2$, then the set $\mathscr{V}(S')$ of tangential fields can be obtained as did in Example 2 of the last section. Denote $\mathscr{M}_1$ the set of all pseudodifferential operators of order 1 with symbol vanishing on $N^*(S_1), N^*(S_2), N^*(\Gamma)$, then $\mathscr{M} \subset \mathscr{M}_1$. Since $\mathscr{M}_1$ can be produced by $\mathscr{V}(S')$, then $\mathscr{M}$ can be produced by $\mathscr{V}(S')$ in the neighborhood of (x, η) for any η. On the other hand, in the neighborhood of $(x_0, \xi^{(1)}$, or $(x_0, \xi^{(2)}$, $\mathscr{M}_1$ coincides with $\mathscr{M}$, then the conditions (1),(2) in Definition 3.6 are satisfied.

Example 3.5. The modulo produced by the operator $M = (1 - \Delta)^{-1/2} D_{x_1} D_{x_2}$ does not satisfy the three points condition. Indeed, the symbol of M is $\xi_1 \xi_2 (1 + \xi^2)^{-1/2}$. Taking $\xi^{(1)} = \xi^{(2)} = \eta = (1, 0)$. If in a neighborhood of $(1, 0)$ the vector field $\alpha_1 D_{x_1} + \alpha_2 D_{x_2}$ can be expressed by M, then $\alpha_1 = 0$, so that Z_1, Z_2 should be chosen as D_{x_2}. However, in any neighborhood of $\eta = (1, 0)$ the operator M cannot be produced by D_{x_2}.

3.5.2 *Pseudo-composition*

It is easy to extend the result in Theorem 3.10 to the case for products of arbitrary functions. To do it we will use the notation $\mathscr{L}(u, M^I u)$, which stands for the function produced by nonlinear composition of its arguments in finite times. Such a notation is called **pseudo-composition** in the sequel.

Lemma 3.14. *Assume that $A \in OpS^0$ with its symbol supported in a small convex cone, $\mathscr{M}$ is an algebra of OpS^0 operators satisfying three points condition, then for any $M \in \mathscr{M}$, $Mf(Au)$ is a pseudo-composition.*

Proof. Since the wave front set of Au is included in Γ, then the wave front set of $f(Au)$ also belongs to Γ. Now let $B(x, D)$ be a pseudodifferential operator of order 0 with its symbol equal to 1 on *supp* $Sym(A)$ and supported in a neighborhood of *supp* $Sym(A)$. Then

$$Mf(Au) = BMf(Au) + (I - B)Mf(Au), \tag{3.79}$$

where the second term is a nonlinear regularized operator, which is automatically pseudo-composition. For any given t, there is a number s, such that

$$\|(I - B)Mf(Au)\|_t \leq (1 + \|u\|_s)^N$$

holds. Then according to the three points condition for $\xi_1 = \xi_2 = \eta$ we have

$$BMf(Au) = \sum C_j(Z_j f(Au)) = \sum C_j(f'(Au)Z_j Au).$$

In view of $Z_j A \in \mathscr{M}$, the first term in the right hand side of Eq. (3.79) is also pseudo-composition. □

Lemma 3.14 can be roughly explained as follows. The action of an operator M on a nonlinear function $f(u)$ can be replaced by a superposition of action of M on u with a pseudo-composition.

Theorem 3.11. *If the modulo $\mathscr{M}$ satisfies the three points condition, $s > n/2$, $F(x, u, \cdots, u_N)$ has compact support with respect to $u_1, \cdots, u_N$, then $u_j(x) \in H^s(\mathscr{M}, k)$ $(1 \leq j \leq N)$ implies $F(x, u_1(x), \cdots, u_N(x)) \in H^s(\mathscr{M}, k)$.*

Proof. Simply denote $F(x, u_1, \cdots, u_N)$ by $F(u)$. When $F \in C_0^\infty$ we have

$$F(u(x)) = \int \exp(i\tau u(x))\hat{F}(\tau)d\tau,$$

where $\hat{F}(\tau)$ is the Fourier transformation of F with respect to u. If $1 = \sum E_j(x, D)$ is a partition of unity, then

$$\exp(i\tau u) = \prod \exp(i\tau E_j U).$$

By using Theorem 3.10 $M \exp(i\tau u)$ can be written as the pseudo-composition $\mathscr{L}(\exp(i\tau E_j u, M_k \exp(i\tau E_j u))$. Meanwhile $M_k \exp(i\tau E_j u)$ is a pseudo-composition according to Lemma 3.13. Hence

$$M \exp(i\tau u) = \mathscr{L}'(\tau u, \tau M_k u),$$

where the form of $\mathscr{L}'$ is independent of τ, so that the right hand side of the above equality is increasing as a polynomial of τ at most. Since $\hat{F}$ is rapidly decreasing, we know the integral

$$\int \mathscr{L}'(\tau u, \tau M_k u)\hat{F}(\tau)d\tau$$

is convergent, and the integral is also in H^s.

The general terms $M^I F(u)$ can be treated by using the same method, then the conclusion $F(u) \in H^s(\mathscr{M}, k)$ can be established. □

3.5.3 *Theorem on interaction of conormal singularities*

Based on the previous preparations we can discuss the interaction of conormal singularities now. Consider the semilinear strictly hyperbolic equation

$$P(x, D_x)u = F(u, \cdots, D^\beta u, \cdots)_{|\beta| \leq m-1}, \tag{3.80}$$

where $P(x, D_x)$ is a strictly hyperbolic operator. Assume that S_1, S_2 are two characteristic surfaces intersecting at Γ, and $S_3, \cdots, S_m$ are other characteristic surfaces issuing from Γ, then we have

Theorem 3.12. *If $u \in H^s$ $(s > n/2 + m)$ is a solution of Eq. (3.80), and $u \in H^{s,k}(R^n_-, S_1 \cup S_2)$ for $x_n < 0$, then the following facts hold in R^n_+*

(1) $u \in H^{s+k}$ outside *$\bigcup_{i=1}^n S_i$;*

(2) $u \in H^{s,k}(\Omega_+, S_i)$ in a neighborhood of $S_i \setminus \Gamma$ (i=1,2);

(3) Denote $t = 2s - n/2 - m + 1$, then $u \in H^{s+k}$ in a neighborhood of $S_j \backslash \Gamma$ $(j = 3, \cdots, m)$ as $s+k \leq t$, otherwise $u \in H^{t,\ell}(\Omega_+, S_j)$ $(j = 3, \cdots, m)$ with $\ell = [s + k - t]$.

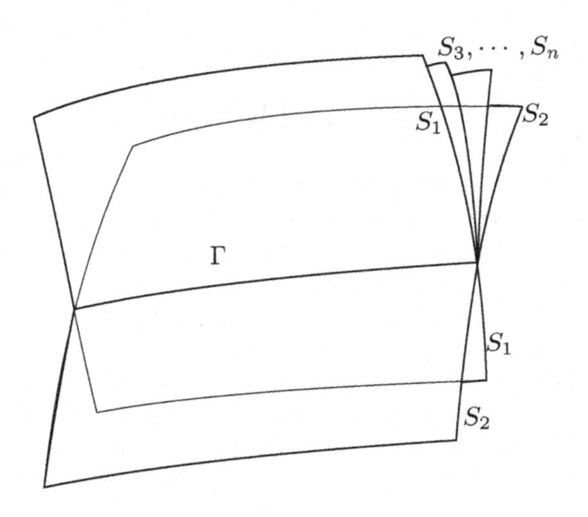

Figure 3.1 Interaction of conormal singularities

Proof. Let $\mathscr{S} = (S_1, \cdots, S_m, \Gamma)$, and denote by $\mathscr{M}$ the modulo of pseudodifferential operators with symbols vanishing on the conormal bundle of all S_i and Γ. Let $H^{s,k}(\mathscr{M})$ be the conormal space defined in the beginning of this section, then $u \in H^{s,k}(\mathscr{M})$ in Ω_-, because $H^{s,k}(\mathscr{M})$ coincides with $H^{s,k}(S_1 \cup S_2)$ in R^n_-. Since $\mathscr{M}$ satisfies the three points condition in R^n_+, then Theorems 3.10 and 3.11 shows that $H^{s,k}(\mathscr{M})$ is closed with respect to nonlinear composition. Therefore, the key point to prove the theorem is to verify the commutation relation. That is, for any $M \in \mathscr{M}$ we have to verify

$$[P, M] \in OpS^{m-1} \cdot \mathscr{M} + OpS^0 \cdot P \tag{3.81}$$

The validity of (3.81) only depends on the principal symbol. Next we verify it in different cases for (x_0, ξ). First, if $x_0 \notin S_i$ for each i, (3.81) is trivial. If $x_0 \in S_i \setminus \Gamma$ for some i, then $Sym([P, M])$ vanishes on $N^*(S_i)$, so that $[P, M]$ can be expressed by $\mathscr{M}$. If $x_0 \in \Gamma$, ξ is not on any $N^*(S_i)$, then (3.81) also holds because any symbol of pseudodifferential operator can be expressed by $\mathscr{M}$ in this case. Hence the only remaining case is $x_0 \in \Gamma, \xi \in N^*(S_i)$ for some i.

In the last case we notice that in a microlocal neighborhood of $T^*(S_i)$ the operator M is equivalent to the tangential vector field of S_i, the $[P, M]$

vanishes on $N^*(S_i)$. On the other hand, since the characteristics of P is simple (it is strictly hyperbolic), then the restriction of $Sym(P)$ on $T^*_\Gamma = \cup_{x\in\Gamma} T^*_x$ only has simple roots on $T^*(\Gamma) \cap T^*(S_i)$. Therefore, we can find a symbol $\lambda(x,\xi)$ of order zero, such that the principal symbol of $[P,M] - \lambda(x,D)P$ vanishes on T^*_Γ. It also means that the principal symbol vanishes on $T^*_\Gamma \cap T^*(S_i)$, so that $[P,M]-\lambda(x,D)P \in \mathscr{M}$ microlocally. That is (3.81).

Having proved the commutation relation Eq. (3.81), we can construct the extended system like we did in the proof of Theorem 3.8 and Theorem 3.9. Acting the element M_i $(i = 1, \cdots, N)$ of $\mathscr{M}$ on Eq. (3.81), we obtain

$$P(M_i u) = [P, M_i]u + M_i F(u, \cdots, D^\beta u),$$

By using the commutator relation

$$[P, M_i] = \sum B_{ij} M_j + B_{i0} + \lambda_i P,$$

$P(M_i u)$ can be written as

$$\sum B_{ij} M_j u + B_{i0} u + \lambda(x,D) F(u, \cdots, D^\beta u) + M_i F(u, \cdots, D^\beta u),$$

where $|\beta| \leq m - 1$. Furthermore, Theorem 3.11 indicates that $M_i F(u, \cdots, D^\beta u)$ is a pseudo-composition, and can be written as $\mathscr{L}(u, \cdots, M_j D^\beta u)$. Then by using paralinearization (see Appendix), we can obtain

$$P(M_i u) = \sum_j D_{ij} M_j u + D_{i0} u + r_i,$$

where D_{ij} is $Op\Sigma^{m-1}_{s-n/2-m+1}$ paradifferential operator, $r_i \in H^{2s-n/2-2m+2}$. Hence we can obtain a system of paradifferential equations

$$PU = DU + R, \tag{3.82}$$

where $U = {}^t(u, M_1 u, \cdots, M_N u)$, P is a matrix of partial differential operators, whose principal symbol is a diagonal matrix with same element in its diagonal, $R \in H^{2s-n/2-2m-2}$. Notice that $U \in H^{s-1}(R^n)$ and $U \in H^s$ as $x_n < 0$. Moreover, U belongs to H^{s-k+1} microlocally outside $T^*(S_1) \cup T^*(S_2)$. By using Theorem 3.4 we know that singularities of U propagate along bicharacteristics, so that U is H^s regular in whole R^n. As for the microlocal regularity outside $T^*(S_1) \cup T^*(S_2)$, U is in H^{t_1}, where

$$t_1 = \min\left(s + k - 1, 2s - \frac{n}{2} - m + 1\right).$$

The commutator relation of higher order can be established by induction. Based on it we can construct a system of paradifferential equations for the vector U_k with $M^I u$ (M is the generators of $\mathscr{M}$, $|I| \le k$) as its elements. This system has the same form as Eq. (3.62). Its principal symbol is a diagonal matrix with same element in its diagonal. Therefore, $M^I u \in H^s$ as $|I| \le k$, i.e. $u \in H^{s,k}(S_i)$. As for the points outside $T^*(S_1) \cup T^*(S_2)$, $M^I u \in H^{\min(s+k-\ell,t)}$ holds microlocally, if $|I| \le \ell$, and $t = 2s-n/2-m+1$. Therefore, by taking $\ell = [s+k-t]$ we obtain the regularity of u near $S_3, \cdots, S_n$ as shown in the conclusion (3) of the theorem. □

3.5.4 *Reflection of conormal singularities*

It is also possible to study the reflection of singularities in the scheme of conormal distributions. When a characteristic surface carrying conormal singularity intersect with boundary transversally, the singularity may propagate along other characteristics issuing from the intersection, so that reflection of singularities occurs. Reflection of singularities can also be considered as that the boundary gives an interaction to the propagation of singularities. To study such a problem one has to overcome some new difficulties, like to distinguish the regularity in normal direction and the regularities in tangential direction, to proceed calculus of pseudodifferential operators in a domain with boundary etc. Here we will only introduce a result given by [14].

Assume that Ω is an open set in R^n, $\Omega_+ = \Omega \cap \{x_n > 0\}$. Consider a boundary value problem

$$\begin{cases} P(x, D_x)u = f(x, u, \cdots, D^{n-2}u), \ x_n > 0; \\ Bu|_{x_n=0} = 0 \end{cases} \tag{3.83}$$

where P is a strictly hyperbolic operator with respect to x_1, the boundary $S_0 : \ x_n = 0$ is non-characteristic, the opeator B satisfies a uniform Lapatinski condition. If Δ is a submanifold of dimension $n-2$ on S_0, while $S_1, \cdots, S_N$ are characteristic surfaces through Δ, intersecting with S_0 transversally. Denote $\mathscr{S} = \{S_0, S_1, \cdots, S_N\}$, $\mathscr{M}$ is a modulo of pseudodifferential operators, whose principal symbol vanishes on the conormal bundle of any surface in $\mathscr{S}$. $H^{s,k} = H^{s,k}(\mathscr{M})$, $H^{s,k}_\Delta$ is the conormal distribution with respect to the submanifold Δ on $x_n = 0$. Then we have the following conclusion:

Theorem 3.13. *If $u \in H^{s+m-1}(\Omega_+)$ is a solution of Eq. (3.83), $s > n/2-1, u \in H^{s+m-1,k}(\Omega_+ \cap \{x_1 < 0\})$, then $u \in H^{s+m-1,k}(\Omega_+)$, and the trace of u on the boundary $\partial\Omega$ satisfies $\gamma u \in H^{s,k}_\Delta(\partial\Omega)$.*

Readers can find the proof of the theorem in [14]. The theorem indicates that the conormal singularities of u at $x_1 < 0$ holds when x_1 becomes positive. Particularly, if the characteristics $S_1, \cdots, S_{N_1}$ through Δ are downward, and the characteristics $S_{N_1+1}, \cdots, S_N$ through Δ are upward, then the theorem indicates the conormal singularity on any surface of $S_1, \cdots, S_{N_1}$ will reflect to all upward surfaces $S_{N_1+1}, \cdots, S_N$.

Chapter 4

Propagation of singularities for fully nonlinear equations

4.1 Theorem of propagation of singularities for principal type equations

This chapter is devoted to the analysis of singularities for fully nonlinear partial differential equations. For fully nonlinear equations both the path of propagation of singularities and the description of the singularities are related to the solution of the equations. Therefore, the discussion on the propagation of singularities in the case for fully nonlinear equations will be more complicated than the discussion in the case for linear or semilinear equations. In order to use the method developed in linear case to nonlinear case people first have to linearize the nonlinear equations. There are many ways to linearize a given nonlinear equation. Among them an efficient method is paralinearization, i.e. linearization by using paradifferential operators. Since such a linearization only neglect the terms with higher regularities, so that it does not influence the description of singularities of an assigned level. Therefore, by such a linearization the problem of propagation of singularities for nonlinear equations will be simply reduced to a linear problem. The theory on paradifferential operators can be found in [19]. To reader's convenience we give a sketch on the main conclusion of this theory in Appendix.

We have given a theorem on propagation of singularities of solutions to linear partial differential equations of principal type and its proof in Chapter 2. In this chapter we will give a new proof by using the method of microlocal energy estimates, because some modification of this new proof can deduce a corresponding proof for nonlinear equations. To avoid to bring readers to a long proof of a mathematical statement, we decompose the theorem to two steps. Such a treatment will also let people be easier to master the crucial points and figure out what is new ingredient in the process passing from linear problems to corresponding nonlinear problems.

First let us recall the theorem of propagation of singularities for linear equations (Theorem 2.9).

Theorem 4.1. *Assume that P is a linear partial differential operator of principal type in strict sense defined in a domain Ω with C^∞ coefficients, u is a real solution of $Pu = f$ with $f \in H^{s+m-1}(\Omega)$, $A_0 = (x_0, \xi_0) \in T^*(\Omega)$ satisfies $p_m(x_0, \xi_0) = 0$. Assume that γ is a bicharacteristic strip of P passing through A_0, then $A_0 \notin WF_s(u)$ implies $\gamma \cap WF_s(u) = \emptyset$.*

The new proof of Theorem 4.1 is based on a microlocal energy estimate. By using this estimate one can successively improve the microlocal regularity on the bicharacteristic strip γ. According to the definition of microlocal regularity of a given function u at (x_0, ξ_0), we know that for any given set U in $T^*(\Omega)$, $u \in H^s(U)$ means $u \in H^s(x, \xi)$ for any $(x, \xi) \in U$. Moreover, one can also define seminorm of u by

$$|u|_{s,U} = \|Mu\|_{H^s}, \tag{4.1}$$

where M is any pseudodifferential operator of order 0 with support in U.

Let us give some simplification of the statement of Theorem 5.1. By multiplying an elliptic operator Q of order $1 - m$ on P, the problem can be reduced to the case, when P is an operator of order 1. Indeed, if the principal symbol $q(x, \xi)$ of Q is a homogeneous function of degree $1 - m$, then multiplying Q on the both sides of $Pu = f$ gives $QPu = Qf$. Since $q(x, \xi)$ does not vanish identically, then H_{qp_m} is parallel to H_{p_m} on the characteristic strip of P. On the other hand, since the action of an elliptic operator of order $1-m$ transforms the WF^{s-m+1} wave front set of f to its WF^s wave front set, then the theorem of propagation of singularities for the operator P of order m can be derived from the case $m = 1$. Besides, we can also only consider a part of the strip γ near the point (x_0, ξ_0), because the conclusion of the theorem for the whole strip γ can be obtained by finite times of extension of the local conclusion.

Assume that V is a conical neighborhood of the point A, U is a conical neighborhood of the strip γ, we can establish the following energy inequality

$$\|K_\delta u\|_\tau \le \delta \|K_\delta Pu\|_\tau + M(|u|_{\tau,V} + |Pu|_{\tau-1,U} + |u|_{\tau-\tau_1,U} + \|u\|_{-N}), \quad \forall u \in C_0^\infty(\Omega_1), \tag{4.2}$$

where $\tau_1 > 0, \Omega_1 \subset \Omega$, N is an arbitrary integer, K_δ is a suitable pseudodifferential operator of order 0 with symbol having the property listed in the following Lemma 4.1. When $\tau = 0$ the above estimate is reduced to

$$\|K_\delta u\|_0 \le \delta \|K_\delta P u\|_0 + M(\delta)(|u|_{0,V} + \|u\|_{-\tau_1}). \tag{4.3}$$

Next we have to prove two assertions. The first one is the conclusion of Theorem 4.1 can be derived from (4.2), and the second one is the validity of (4.2). As the first step we prove (4.3).

Lemma 4.1. *There is a pseudodifferential operator K_δ with symbol k_δ satisfying the following conditions:*

(1) C^∞ smooth, positive on γ, and having support in U.

(2) homogeneous function of ξ with degree 0.

(3) $Hk_\delta \ge \frac{2}{\delta} k_\delta$ outside V, where H is the Hamilton vector field of $p_m(x,\xi)$.

Proof. Project all elements (x_0, ξ_0), H, γ, U, V on the sphere bundle – a subbundle of cotangent bundle $T^*(\Omega)$, we can construct k_δ on the sphere bundle and then extend all related functions to the whole cotangent bundle as homogeneous functions of ξ with degree 0. Denote the coordinates of the sphere bundle by $y_1, \cdots, y_{2n-1}$. Since the operator P is principal type in strict sense, then $H \ne 0$. Hence one can make a transformation to straighten the bicharacteristics γ and let H become $H_1 = \partial_{y_1}$. Next, without loss of generality we may assume that the image of A_0 is $(1, 0, \cdots, 0)$, the image $\tilde{\gamma}$ of an assigned part of γ is an interval from $(0, 0, \cdots, 0)$ to $(1, 0, \cdots, 0)$. Then we can construct a function $\zeta(y_1)$, such that the support of ζ locates on $(-\delta, 1+\delta)$ for sufficiently small number $\delta > 0$, and the derivative $\zeta'(y_1) > 0$ outside $(1-\delta, 1+\delta)$. Make another C_0^∞ function $\theta_2(y_2, \cdots, y_{2n-1})$ supported on a δ neighborhood of the origin, and let $\phi(y) = \zeta_1(y_1)\theta_2(y_2, \cdots, y_{2n-1})$, then the support of the function $\phi(y)$ contains in a neighborhood $\tilde{U}$ of the image $\tilde{\gamma}$ of γ, and $H_1\phi \ge 0$ on $\tilde{U} \backslash \tilde{V}$.

Now let $\tilde{k_\delta} = \phi \exp(2y_1/\delta)$, then

$$H_1 \tilde{k_\delta} = (H_1\phi)\exp(2y_1/\delta) + \frac{2}{\delta}\tilde{k_\delta} \le \frac{2}{\delta}\tilde{k_\delta}. \tag{4.4}$$

Return to the original coordinates in the cotangent bundle, we proved the lemma. □

Lemma 4.2. *Assume that K_δ is the pseudodifferential operator with the symbol k_δ constructed in Lemma 4.1, then the equality (4.3) holds.*

Proof. Consider the inner product $(K_\delta P u, K_\delta u)$, we have

$$Im(K_\delta P u, K_\delta u) - Im(P K_\delta u, K_\delta u) = Re(iK_\delta^*[P, K_\delta]u, u). \tag{4.5}$$

Let us estimate the two terms in the left-hand side. Obviously,

$$|Im(K_\delta Pu, K_\delta u)| \leq \|K_\delta Pu\|_0 \|K_\delta u\|_0. \tag{4.6}$$

In view of that $P - P^*$ is a pseudodifferential operator of order 0, then

$$|Im(PK_\delta u, K_\delta u)| = |((P - P^*)K_\delta u, K_\delta u)|/2 \leq C\|K_\delta u\|_0^2. \tag{4.7}$$

On the other hand, the operator $[P, K_\delta]$ on the right-hand side of (4.5) is also a pseudodifferential operator of order 0 with the symbol Hk_δ/i, hence the principal symbol of operator $iK_\delta^*[P, K_\delta]$ is $\bar{k}_\delta Hk_\delta$, which is larger than or equal to $2|k_\delta|^2/\delta$. Hence we have

$$iK_\delta^*[P, K_\delta] = \frac{2}{\delta}K_\delta^* K_\delta + S_1 + S_2, \tag{4.8}$$

where S_1, S_2 are pseudodifferential operators of order 0. Moreover, the principal symbol of S_1 is non-negative, the principal symbol of S_2 is 0 outside of V. Then, by using Gårding inequality, we have

$$Re(iK_\delta^*[P, K_\delta]u, u) \geq \frac{2}{\delta}\|K_\delta u\|_0^2 - M\|u\|_{-N}^2 - M|u|_{0,V}^2. \tag{4.9}$$

Taking δ sufficiently small, and substituting Eqs. (4.6), (4.7), (4.9) into Eq. (4.5), we obtain Eq. (4.3). □

Lemma 4.3. *Under the assumptions of Lemma 4.2 the estimate (4.2) holds.*

Proof. Making a properly supported psuododifferential operator E of order τ, whose symbol is $(1 + |\xi|^2)^{\tau/2}$ outside a neighborhood of U, then according to the psuododifferential calculus we know

$$\begin{aligned} \|K_\delta u\|_\tau &= \|\Lambda^\tau K_\delta u\|_0 \\ &\leq \|EK_\delta u\|_0 + \|(\Lambda^\tau - E)K_\delta u\|_0 \\ &\leq \|K_\delta Eu\|_0 + \|[E, K_\delta]u\|_0 + \|(\Lambda^\tau - E)K_\delta u\|_0 \\ &\leq \|K_\delta Eu\|_0 + C|u|_{\tau-1,U} + C\|u\|_{-N}, \end{aligned} \tag{4.10}$$

where N is an arbitrary large number. Equation (4.3) indicates

$$\begin{aligned} \|K_\delta Eu\|_0 &\leq \delta\|K_\delta PEu\|_0 + M(|Eu|_{0,V} + \|Eu\|_{-\tau_1}) \\ &\leq \delta\|K_\delta PEu\|_0 + M(|u|_{\tau,V} + \|u\|_{-N} + |u|_{\tau-\tau_1,U}. \end{aligned} \tag{4.11}$$

Writing $K_\delta PE$ as

$$EK_\delta P + [K_\delta, E]P + [K_\delta, [P, E]] + [P, E]K_\delta$$

and estimating each term, we obtain

$$\begin{aligned} \|K_\delta PEu\|_0 &\leq C(\|K_\delta Pu\|_\tau + \|K_\delta u\|_\tau) \\ &\quad + M(Pu|_{\tau-1,U} + |u|_{\tau-\tau_1,U} + \|u\|_{\tau-1}). \end{aligned} \tag{4.12}$$

Substituting it into Eq. (4.10) and taking δ sufficiently small we obtain Eq. (4.2). □

Proof of Theorem 4.1. Without loss of generality we may assume that the support of u is compact. Here we should indicate that the boundedness of all terms in Eq. (4.2) implies $u \in H^\tau(\gamma)$. Denoting J_η as a mollifier, we have $J_\eta^2 u \in C_0^\infty(\Omega)$. Therefore,

$$\begin{aligned}\|K_\delta J_\eta^2 u\|_\tau \leq \delta \|K_\delta P J_\eta^2 u\|_\tau + M(|J_\eta^2 u|_{\tau,V} + |PJ_\eta^2 u|_{\tau-1,U} \\ + |J_\eta^2 u|_{\tau-\tau_1,U} + \|J_\eta^2 u\|_{\tau-1}).\end{aligned} \tag{4.13}$$

Since $u \in H^{\tau-1}$, and u belongs to $H^\tau(V) \cap H^{\tau-\tau_1}(U)$ microlocally, then

$$|J_\eta^2 u|_{\tau-\tau_1,U} \to |u|_{\tau-\tau_1,U}, \quad |J_\eta^2 u|_{\tau,V} \to |u|_{\tau,V},$$

$$\|J_\eta^2 u|_{\tau-1} \to \|u\|_{\tau-1},$$

if $\eta \to 0$. On the other hand, since $[P, J_\eta^2]$ is a bounded operator from L^2 to L^2, and its norm is bounded uniformly with respect to η, then

$$|PJ_\eta^2 u|_{\tau-1,U} \leq |Pu|_{\tau-1,U} + C|u|_{\tau-1,U},$$

$$\begin{aligned}\|K_\delta P J_\eta^2 u\|_\tau &\leq \|K_\delta J_\eta^2 P u\|_\tau + \|[K_\delta, [J_\eta^2, P]]u\|_\tau + \|[J_\eta^2, P]K_\delta u\|_\tau \\ &\leq \|K_\delta J_\eta^2 P u\|_\tau + \|[K_\delta, [J_\eta^2, P]]u\|_\tau + \|[J_\eta[J_\eta, P]]K_\delta u\|_\tau \\ &\quad + 2\|[J_\eta, P]J_\eta K_\delta u\|_\tau.\end{aligned}$$

The terms are dominated by

$$C(\|Pu\|_\tau + \|u\|_{\tau-1} + \|K_\delta J_\eta u\|_\tau).$$

Hence for sufficiently small η we have

$$\begin{aligned}\|K_\delta J_\eta u\|_\tau &\leq \|K_\delta J_\eta^2 u\|_\tau + \|[K_\delta, J_\eta - I]J_\eta u\|_\tau + \|(J_\eta - I)K_\delta J_\eta u\|_\tau \\ &\leq C_1 \delta \|K_\delta J_\eta u\|_\tau + \frac{1}{2}\|K_\delta J_\eta u\|_\tau + C_2,\end{aligned}$$

where C_2 depends on $\|Pu\|_\tau, \|u\|_{\tau-1}, |u|_{\tau,V}, |u|_{\tau-\tau_1,U}$. Hence, by taking $\delta < 1/(4C_1)$, we obtain

$$\|K_\delta J_\eta u\|_\tau \leq 2C_2. \tag{4.14}$$

Moreover, notice that $J_\eta u \to u$ $(H^{\tau-1})$, as $\eta \to 0$, we then obtain

$$K_\delta J_\eta u \to K_\delta u.$$

By using Banach-Saks theorem and Eq. (4.14) we know $K_\delta u \in H^\tau$. Since k_δ is positive on γ, we know u belongs to $H^\tau(\gamma)$ microlocally.

The above demonstration indicates that $u \in H_\gamma^{\tau-\tau_1} \cap H^{-N}(\Omega)$, $u \in H^\tau_{(A_0)}$ and $Pu \in H^\tau_\gamma$ imply $u \in H^\tau_\gamma$. Since u is assumed to have a compact support, then there is a number $-N$, such that $u \in H^{-N}$. Hence we can improve the regularity of u on γ successively, and finally confirm that the regularity of u is determined by its regularity at A_0 and the regularity of Pu on γ as shown is the theorem. $\square$

Turn to the study on the nonlinear case, we discuss the fully nonlinear equation

$$\mathscr{F}[u] \triangleq F(x, u(x), \cdots, \partial^\alpha u(x), \cdots)_{|\alpha|\leq m} = 0, \tag{4.15}$$

where F is a C^∞ function of its arguments. When $u \in H^s$ and $s > n/2+m$, $\mathscr{F}[u]$ is a continuous function of x. As mentioned above, the characteristic polynomial is

$$p_m(x,\xi) = \sum_{|\alpha|=m} \frac{\partial F}{\partial u_\alpha}(x, u(x), \cdots)(i\xi)^\alpha,$$

which depends on the solution u, so is the bicharacteristic strip. Hence the general form of the theorem of singularity propagation for the fully nonlinear equations is

Theorem 4.2. *Assume that u is a solution of the equation Eq. (4.15),* $u \in H^s_{loc}(\Omega)$, $s > \dfrac{n}{2}+m+2$, $A_0(x_0,\xi_0)$ *is a characteristic point of the equation,* γ *is a bicharacteristic strip passing through* A_0. *Moreover,* $u \in H^t(x_0,\xi_0)$ *with* $t \leq 2s - \dfrac{n}{2} - m - 1$, *then* $u \in H^t_\gamma$.

The proof can be proceeded along the outline of the proof of Theorem 4.1. Here let us emphasize the difference. First, in the nonlinear case we have to assume that u has some regularity initially, then successively to improve its regularity. For instance, we may assume that $u \in H^{m+2}$, then the corresponding Hamilton-Jacobi system has C^1 coefficients, so that the bicharacteristic strip can be well defined. Second, since the coefficients of the characteristic equation is only finitely smooth, then the proof of Theorem 4.1, including the construction of the operator K_δ should be modified.

Like the proof for Theorem 4.1, we first linearize the equation Eq. (4.15). By using the theory of paralinearization and paradifferential operators, the equation can be linearized as

$$\sum_{|\alpha|\leq m} T_{\frac{\partial F}{\partial u_\alpha}} \partial^\alpha u = r, \tag{4.16}$$

where r is a function with higher regularity. $r \in H^{2(s-m)-n/2}$ if $t > 2s - n/2 - m - 1$, and $r \in H^{t-m+1}$ if $t \leq 2s - n/2 - m - 1$. Then the proof is complete provided the following proposition holds.

Theorem 4.3. *Assume $P \in Op(\Sigma_\sigma^m)(\Omega)$, where $\sigma > 1$ is a non-integer, the principal symbol p_m of P is a C^2 real function. Assume also that $u \in H^s, Pu \in H^{t-m+1}(\Omega)$ and $u \in H^t(x_0, \xi_0)$, where s and t satisfy the requirement in Theorem 4.2, then u belongs to $H^t(\gamma)$ microlocally, where γ is a bycharacteristic strip passing through (x_0, ξ_0).*

Proof. The main steps of the proof are basically the same as that for Theorem 4.1. However, since the principal symbol p_m is only a C^2 function, then the function k_δ constructed in Lemma 4.1 is not a C^∞ function. Therefore, we introduce a $C_0^\infty(\Omega)$ function $j_\epsilon(z)$, satisfying $\displaystyle\int j_\epsilon dz$, and replacing k_δ by $k_{\delta\epsilon} = k_\delta * j_\epsilon = \displaystyle\int k_\delta(z-z')j_\epsilon(z')dz'$. Obviously, $k_\delta * j_\epsilon$ is a positive C^∞ function, and

$$Hk_{\delta\epsilon} - \int Hk_\delta(z-z')j_\epsilon(z')dz' \geq \frac{2}{\delta}k_{\delta\epsilon}. \tag{4.17}$$

Making a homogeneous extension of degree 0 with respect to ξ, the symbol then satisfies all conditions in Lemma 4.1. Afterwards, we obtain the operator $K_{\delta\epsilon}$.

Similar to the proof of Lemma 4.2 and Lemma 4.3 with replacing the pseudodifferential operators by corresponding paradifferential operators we can complete the proof of Theorem 4.3. In fact, the rule of the calculus for paradifferential operators and the rule of the calculus for pseudodifferential operators are essential the same, then the main idea of these two lemmas are still available. The operator S_2 in Eq. (4.8) should be replaced by a paradifferential operator of order 1, whose symbol vanishes outside V. To estimate (S_2u, u), we make an operator L, whose symbol $\ell \in S_0$ vanishes outside of V and is equal to 1 on supp$Sym(S_2)$, then $R_2 = (I - L)S_2$ is an operator with lower order. On the other hand, the conjugate operator L^* of the operator L can be expressed as a sum of a C^∞ regular operator and an operator L', whose symbol vanishes outside of V. Denote by R_2 the paradifferential operator of lower order produced in the process of the calculus of paradifferential operators, then for $\tau_1 < \dfrac{1}{2}$ we have

$$\begin{aligned}
|(S_2u,u)| &\leq |(LS_2u,u)| + |(R_2u,u)| \\
&= |(S_2u,L^*u)| + |(R_2u,u)| \\
&\leq |(S_2u,L'u)| + |(S_2u,R_1u)| + |(R_2u,u)| \\
&\leq M(|u|_{0,V}^2 + \|u\|_{-\tau_1}^2).
\end{aligned} \tag{4.18}$$

Hence Eq. (4.9) still holds. □

Remark 4.1. If the nonlinearity of the equation Eq. (4.15) is weaker, then the conclusion of Theorem 4.2 can be better. For instance, if Eq. (4.15) is a quasilinear equation, then the conclusion of the theorem holds for $s > n/2+m+1, t \leq 2s-n/2-m$. Moreover, if Eq. (4.15) is semilinear, then the conclusion of the theorem holds for $s > n/2+m-1, t \leq 2s-n/2-m+1$.

From Theorem 4.2 we obtain the following theorem on extension of H^s regularity of solutions to nonlinear equations.

Theorem 4.4. *Assume that Eq. (4.15) is a hyperbolic equation with respect to* x_n*,* u *is its* $H^s(\Omega)$ *solution with* $s > \frac{n}{2} + m + 2$*,* $\Omega \cap \{x_n > 0\}$ *locates in the determinacy region of* $\Omega_- = \Omega \cap \{x_n < 0\}$*. If* $u \in H^t$ *with* $t > s$ *in* $x_n < 0$*, then* $u \in H^t$ *in the whole domain* Ω*.*

Proof. Taking $s_1 = \min\{2s - n/2 - m - 1, t\}$, for any point x_0 in $\Omega \cap \{x_n > 0\}$ and any direction ξ_0 we consider the regularity of u at (x_0, ξ_0). If (x_0, ξ_0) is not any characteristic point of Eq. (4.15), then the equation is microlocally elliptic, so that we know $u \in H^{2s-n/2-m-1}$ by using the regularity theorem for elliptic equation. This implies $u \in H^{s_1}(x_0, \xi_0)$. If (x_0, ξ_0) is a characteristics point of Eq. (4.15), then Theorem 4.2 implies $u \in H^{s_1}(x_0, \xi_0)$. Since ξ_0 can be any direction in R^n_ξ, then $u \in H^{s_1}(x_0)$. Due to the arbitrariness of x_0 we know $u \in H^{s_1}(\Omega)$. Replacing s by s_1, we can further improve the regularity of u by using the same method. Finally, we obtain $u \in H^s(\Omega)$. □

The reflection of singularities is also extensively studied. The difficulty is that the original paralinearization does not work in the domain with boundary. The concept of tangential paralinearization and the corresponding calculus was established in [122]. In this case both the description of singularities and the reduction of the equation in tangential direction and in normal direction of the boundary should be treated by using different method. The problem of singularity reflection of nonlinear equation in the case when the bicharacteristics bearing singularities transversally intersects boundary is discussed in [122], while in the case when the bicharacteristics is tangential to the boundary is discussed in [145].

4.2 Propagation of conormal singularities for nonlinear equations

In Section 4.1 we discussed the propagation of singularities described by wave front sets for fully nonlinear equations. Like the semilinear equation case, in the framework of wave front sets one can only study the propagation of singularities with order no more that $2s$ for general nonlinear equations. Hence in the sequel we will discuss the propagation of conormal singularities for fully nonlinear equations. As mentioned in Chapter 3, the description of conormal singularities of a given distribution u depends on a variety of submanifolds $\mathscr{S}$. The regularities of u on the normal direction and the tangential direction of submanifolds in $\mathscr{S}$ are different. If u is a solution of a given partial differential equation, S is a hypersurface bearing some singularities of u, then S must be a characteristics of the equation. Since the equation is fully nonlinear, the characteristic hypersurface S itself may also be singular in some sense. In accordance, the statement as well as the proof of the related propositions will be more complicated. In this section we will mainly discuss the propagation of single conormal wave for nonlinear hyperbolic equations.

Again assume that Ω is a domain in R^n, intersecting with $x = 0$.

$$F(x, u(x), \cdots, \partial^\alpha u(x), \cdots)_{|\alpha|\le m} = 0 \tag{4.19}$$

in a nonlinear equation given in Ω. Let u be the H^{s+m} solution with $s > n/2$. Denote by $u^{(\alpha)}$ the derivative $\partial^\alpha u(x)$, and make the linearized operator P for the nonlinear operator F at u as

$$P = \sum_{|\alpha|\le m} \frac{\partial F}{\partial u^{(\alpha)}}(x, u(x), \cdots)\partial_x^\alpha. \tag{4.20}$$

Assume that for given u the operator P is strictly hyperbolic with respect to x_n, and assume that Ω is included in the determinacy region of $\Omega_- = \Omega \cap R^n_-$. Denote $x' = (x_2, \cdots, x_{n-1})$, and assume that $S : x_1 = \phi(x', x_n)$ is a characteristic surface of Eq. (4.19), then we are going to confirm that the conormal regularity of u in Ω_- implies its conormal regularity in the whole region Ω.

Theorem 4.5. *Assume that* $s > \dfrac{n}{2} + \dfrac{7}{2}$, $\sigma > \dfrac{n}{2} + \dfrac{3}{2}$, $u \in H^{s+m}(\Omega)$ *is a solution of Eq. (4.19),* S *is an* H^σ *regular characteristic surface. If* $S \in C^\infty$ *and* $u \in H^{s+m,\infty}(S)$ *in* Ω_-*, then* S *is* C^∞ *in the whole* Ω *and* $u \in H^{s+m,\infty}(S)$.

The proof of this theorem will be proceeded by several Lemmas. First, to overcome the difficulty to give a proper definition of conormal singularity we make a coordinate transformation χ to straighten S to $x_1 = 0$. Since in the new coordinate system the conormal singularity is well defined, then the solution of the original equation is regarded as a composition of a distribution u with a specific conormal singularity and a non-smooth coordinate transformation. When we make paralinearization for nonlinear partial differential equations such a composition also plays important role. The composition is called **paracomposition**, which was established in [3]. The main results on paracomposition will be briefly described in Appendix of this book. By using the theory of paradifferential operators and paracomposition we establish a differential equation in the new coordinate system for the distribution $\chi^* u$, then by using the bootstrap way to improve the regularity of χ and $\chi^* u$ repeatedly, and finally obtain the required result.

Let $x' = (x_2, \cdots, x_{n-1})$ and let $x_1 = \phi(x', x_n)$ be the characteristic surface of (4.15), then ϕ satisfies the equation

$$\frac{\partial \phi}{\partial x_n} = \lambda(\phi, x', x_n, u(x), \cdots, -1, \nabla_{x'}\phi), \tag{4.21}$$

where $\lambda(x_1, x', u, \cdots, \xi_1, \xi')$ is a real root of the equation

$$\sum_{|\alpha|=m} \frac{\partial F}{\partial u^{(\alpha)}}(x, u, \cdots)\xi^\alpha = 0 \tag{4.22}$$

with respect to ξ_n.

Denote the original coordinates by $(\bar{x}_1, \cdots, \bar{x}_n)$, we introduce an H^s homeomorphism χ:

$$\bar{x}_1 = x_1 + \phi(x', x_n), \ \bar{x}' = x', \ \bar{x}_n = x_n. \tag{4.23}$$

Then the equation of the characteristic surface S in the new coordinate system is $x_1 = 0$. The function $v(\phi(x', x_n), x', x_n)$ can be regarded as the value of $v \circ \chi$ on $x_1 = 0$.

Lemma 4.4. *Assume that* $s > \frac{n}{2} + \frac{7}{2}, \sigma > \frac{n}{2} + \frac{3}{2}$, $u \in H^{s+m}(\Omega)$, ϕ *satisfies Eq. (4.21), then* $\phi \in H^{s-1/2}$. *Furthermore, if* $\chi^* u \in H^{s+m,k}$, *then* $\phi \in H^{s+m-1/2}$.

Proof. First let us derive the paralinearized equation. When Eq. (4.21) is regarded as a nonlinear partial differential equation for ϕ, the function $u(x_1, x' x_n)$ should be regarded as a given function. Since u is only H^{s+m} regular, then in the process of paralinearization for Eq. (4.21) we must consider the influence of nonlineararity and nonsmoothness simultaneously.

Without loss generality we assume $\sigma < s - 1/2$, because the conclusion of the lemma automatically holds otherwise. Since $\chi \in H^{\sigma}$, then the imbedding theorem implies $u^{(\alpha)} \circ \chi \in C^{\sigma - n/2}$, $(\partial_{x_1} u^{(\alpha)}) \circ \chi \in C^{\sigma_1}$, where $\sigma_1 = \min(s - n/2 - 1, \sigma - n/2)$. From the rule of paracomposition Eq. (A.32) we know that for any function $v \in H^{s'}$ with $s' > n/2 + 1$,

$$v \circ \chi = \chi^* v + T_{v' \circ \chi} \chi + R, \quad R \in H^{\sigma + \sigma_1}. \tag{4.24}$$

Therefore,

$$\begin{aligned} u^{(\alpha)}(\phi, x', x_n) &= u^{(\alpha)} \circ \chi\big|_{x_1=0} \\ &= \chi^* u^{(\alpha)}\big|_{x_1=0} + (T_{(\partial_{x_1} u^{(\alpha)}) \circ \chi} \chi)_{x_1=0} + R_1\big|_{x_1=0} \\ &= \chi^* u^{(\alpha)}\big|_{x_1=0} + T_{(\partial_{x_1} u^{(\alpha)})(\phi, x', x_n)} \phi + R_1\big|_{x_1=0} + Q_1 \phi, \end{aligned} \tag{4.25}$$

where Q_1 is a σ_1-regular operator. Furthermore, the trace theorem implies $R_1\big|_{x_1=0} \in H^{\sigma + \sigma_1 - 1/2}$, then

$$u^{(\alpha)}(\phi, x', x_n) \in H^{(\sigma)}.$$

Similarly,

$$\begin{aligned} (\partial_{x_1} u^{(\alpha)})(\phi, x', x_n) = (\chi^* \partial_{x_1} u^{(\alpha)})\big|_{x_1=0} &+ T_{(\partial_{x_1}^2 u^{(\alpha)})(\phi, x', x_n)} \phi \\ + R_2\big|_{x_1=0} + Q_2 \phi, \end{aligned}$$

where $R_2\big|_{x_1=0} \in H^{s-3/2}$, Q_2 is a σ_2-regular operator with $\sigma_2 = \min(s - n/2 - 2, \sigma - n/2)$. Moreover, $(\chi^* \partial_{x_1} u^{(\alpha)})\big|_{x_1=0} \in H^{s-3/2}$, then

$$(\partial_{x_1} u^{(\alpha)})(\phi, x', x_n) \in H^{\min(s-3/2, \sigma)} \subset H^{\sigma - 1}. \tag{4.26}$$

Paralinearizing the right-hand side of Eq. (4.21), we have

$$\lambda(\phi, x', x_n, u^{(\alpha)}; -1, \phi_{x'}) = T_{\frac{\partial \lambda}{\partial \phi}} \phi + \sum T_{\frac{\partial \lambda}{\partial u^{(\alpha)}}} u^{(\alpha)}(\phi, x', x_n) + T_{\frac{\partial \lambda}{\partial \xi'}} \phi_{x'} + R_3, \tag{4.27}$$

where $R_3 \in H^{2(\sigma-1)-(n-1)/2}$. By using Eq. (4.25) $T_{\frac{\partial \lambda}{\partial u^{(\alpha)}}} u^{(\alpha)}(\phi, x', x_n)$ can be written as

$$\begin{aligned} T_{\frac{\partial \lambda}{\partial u^{(\alpha)}}} T_{\partial_{x_1} u^{(\alpha)}} \phi &+ T_{\frac{\partial \lambda}{\partial u^{(\alpha)}}} (\chi^* u^{(\alpha)}\big|_{x_1=0} + R_1\big|_{x_1=0} + Q_1 \phi) \\ &= T_{\frac{\partial \lambda}{\partial u^{(\alpha)}} \cdot \partial_{x_1} u^{(\alpha)}} \phi + R_4 + R_5, \end{aligned} \tag{4.28}$$

where

$$R_4 = T_{\frac{\partial \lambda}{\partial u^{(\alpha)}}} (\chi^* u^{(\alpha)}\big|_{x_1=0} + R_1\big|_{x_1=0} + Q_1 \phi) \in H^{\min(s-1/2, \sigma + \sigma_1 - 1/2)},$$

and R_5 is the remainder produced in the composition of paraproduct, so that is an $H^{2\sigma-(n+1)/2}$ function. Applying all these estimates in the equation of ϕ, we have

$$\phi_{x_n} = T_{\frac{\partial\lambda}{\partial\xi'}}\phi_{x'} + T_A\phi + R_6, \tag{4.29}$$

where $\dfrac{\partial\lambda}{\partial\xi'}, A \in H^{\sigma-1}$, R_6 is the sum of R_3, R_4, R_5. Therefore, denoting

$$\begin{aligned}\theta &= \min\left(2(\sigma-1) - \frac{n-1}{2}, s - \frac{1}{2}, \sigma+\sigma_1 - \frac{1}{2}\right)\\ &= \min\left(2\sigma - \frac{n}{2} - \frac{3}{2}, s - \frac{1}{2}\right),\end{aligned}$$

hence $\phi \in H^\theta$. Indeed, because $\sigma - 1$ is the regularity index of $\dfrac{\partial\lambda}{\partial\xi'}$ and A, and $n-1$ is the number of independent variables in Eq. (4.29), we have $\sigma > n/2 + 3/2$, i.e. $(\sigma-1) - (n-1)/2 > 1$, then from Theorem 4.3 we know that the H^θ regularity propagates along the bicharacteristic strip of Eq. (4.29). Therefore, ϕ is an H^θ function in $t > 0$. Now if $\theta = s - 1/2$, then the conclusion in the first part has been proved. Otherwise, we may repeat the process with replacing σ by θ to improve the regularity of ϕ successively, and finally reach $\phi \in H^{s-1/2}$.

If $\chi^* u$ has higher regularity, ϕ can also be more regular. Let $\sigma = s-1/2$, in order to obtain the conclusion of this lemma, we only have to prove that $\phi \in H^{s+k-1/2}$ and $\chi^* u \in H^{r+m,k+1}$ imply $\phi \in H^{s+k+1/2}$.

The remains of the proof is similar to the first part. Since χ is an $H^{\sigma+k}$ transformation, then the regularity of the remainders should be

$$\begin{aligned}&R \in H^{\sigma+k+\theta} \quad \text{with} \quad \theta = \min(\sigma+k-n/2, s-n/2-1),\\ &R_1\big|_{x_1=0} + Q_1\phi \in H^{\sigma+k+(s-n/2-1)-1/2} \subset H^{s+k+1/2},\\ &R_2\big|_{x_1=0} + Q_2\phi \in H^{\sigma+k+(s-n/2-2)-1/2} \subset H^{s+k+1/2}.\end{aligned}$$

To obtain better regularity of $\chi^* u^{(\alpha)}$ and $\chi^*\partial_{x_1} u^{(\alpha)}$ we use the theorem of superposition of paracomposition and paradifferential operator (see Theorem A.28 in Appendix) to obtain

$$\chi^* u^{(\alpha)} = T_{\ell^{(\alpha)}}\chi^* u + R_\alpha u, \tag{4.30}$$

where $T_{\ell^{(\alpha)}} \in \Sigma^{|\alpha|}_{\sigma+k-n/2-1}$, R_α is a $(\sigma+k-n/2-1-|\alpha|)$ regular operator, then

$$\begin{aligned}&\chi^* u^{(\alpha)} \in H^{s+m-|\alpha|,k+1} \subset H^{s,k+1},\\ &\chi^*\partial_{x_1} u^{(\alpha)} \in H^{s-1,k+1}.\end{aligned}$$

Then we have

$$\chi^* u^{(\alpha)}\big|_{x_1=0} \in H^{s+k+1/2}, \ (\chi^* \partial_{x_1} u^{(\alpha)})\big|_{x_1=0} \in H^{s+k-1/2},$$

$$u^{(\alpha)}(\phi, x', x_n) \in H^{s+k+1/2}, \ \partial_{x_1} u^{(\alpha)}(\phi, x', x_n) \in H^{s+k-1/2}.$$

On the other hand, the remainders produced in the process of paralinearization satisfy

$$R_3 \in H^{2(\sigma+k-1)-(n-1)/2},$$

$$R_4 \in H^{s+k+1/2},$$

$$R_5 \in H^{s+k+(\sigma+k+(n-1)/2)}.$$

Then from $\sigma = s - 1/2 > n/2 + 3$ we know $R_3 + R_4 + R_5 \in H^{s+k+1/2}$. According to Theorem 4.3 we have $\phi \in H^{s+k+1/2}$. As mentioned above, we are then led to the second conclusion of the theorem by induction. □

Lemma 4.4 implies that the better regularity of $\chi^* u$ implies better regularity of ϕ. Next we look for the paradifferential equation satisfying by $\chi^* u$ to get more regularity of $\chi^* u$.

Lemma 4.5. *Assume $s > n/2 + 7/2, \sigma = s - 1/2, \phi \in H^{s+k-1/2}, u \in H^{s+m}, \chi^* u \in H^{s+m,k}$, then $\chi^* u$ satisfies*

$$T_{p^*} \chi^* u = R, \tag{4.31}$$

where $p^* = p^*_m + p^*_{m-1}$, p^*_{m-j} is the symbol of a differential operator of order $m - j$ with coefficients in $C^{m-j-n/2-1,k}$, $R \in H^{s+1,k+1}$.

Proof. As indicated in Lemma 4.4, $\chi^* u \in H^{s+m,k}$ implies $\chi^* u^{(\alpha)} \in H^{s,k}$, then by using Theorem A.26 in Appendix the term $\chi^*(F(x, u, \cdots, u^{(\alpha)}, \cdots))$ can be paralinearized as

$$\sum_\alpha T_{\frac{\partial F}{\partial u^{(\alpha)}}} \chi^* u^{(\alpha)} + R_1 = 0, \tag{4.32}$$

where $R_1 \in H^{(s-1/2)+(s-n/2-1),k} \subset H^{s+1,k+1}$, The argument of $\dfrac{\partial F}{\partial u^{(\alpha)}}$ is

$$u^{(\alpha)} \circ \chi = \chi^* u^{(\alpha)} + T_{\nabla(u^{(\alpha)}) \circ \chi} \chi = R_2 \in H^{s-1/2,k}.$$

In the left-hand side of Eq. (4.32), for the terms with $|\alpha| \le m - 2$, we have

$$\frac{\partial F}{\partial u^{(\alpha)}} \in H^{s-1/2,k} \subset C^{s-(n+1)/2,k}.$$

Then Theorem A.14 in Appendix implies

$$T_{\frac{\partial F}{\partial u^{(\alpha)}}}\chi^* u^{(\alpha)} \in H^{s+m-|\alpha|,k} \subset H^{s+1,k+1}.$$

For the terms with $|\alpha| = m$ or $m-1$ in the left-hand side of Eq. (4.32) we write $\chi^* u^{(\alpha)}$ as $T_{\ell^{(\alpha)}}\chi^* u + R_\alpha u$, where $\ell^{(\alpha)} \in \Sigma^{|\alpha|}_{\sigma+k-1/2-1}$, R_α is a $(\sigma+k-n/2-1-|\alpha|)$ regular operator with $\sigma = s+k-1/2$. From $u \in H^{s+m}$ we know $R_\sigma u \in H^{s+1,k+1}$. Moreover, denoting by $\ell^{(\alpha)}_j$ the homogeneous terms of degree j, we have

$$T_{\ell^{(\alpha)}}\chi^* u - T_{\ell^{(\alpha)}_m}\chi^* u + T_{\ell^{(\alpha)}_{m-1}}\chi^* u \in H^{s+m,k} \subset H^{s+1,k+1},$$

which can be regarded as reminder. Hence, Eq. (4.32) can be written as

$$\sum_{|\alpha|=m} T_{\frac{\partial F}{\partial u^{(\alpha)}}} T_{\ell^{(\alpha)}_m}\chi^* u + \sum_{|\alpha|=m} T_{\frac{\partial F}{\partial u^{(\alpha)}}} T_{\ell^{(\alpha)}_{m-1}}\chi^* u + \sum_{|\alpha|=m-1} T_{\frac{\partial F}{\partial u^{(\alpha)}}} T_{\ell^{(\alpha)}_{m-1}}\chi^* u = R_2, \tag{4.33}$$

where $R_2 \in H^{s+1,k+1}$. Since the coefficients of $\ell^{(\alpha)}_m$ belongs to $C^{s+k-(n+3)/2}$, the coefficients of $\ell^{(\alpha)}_{m-1}$ belongs to $C^{s+k-(n+5)/2}$, and $\dfrac{\partial F}{\partial u^{(\alpha)}} \in C^{s-(n+1)/2,k}$, then according to the property of paradifferential operators we have

$$T_{\frac{\partial F}{\partial u^{(\alpha)}}} T_{\ell^{(\alpha)}_m}\chi^* u = T_{\frac{\partial F}{\partial u^{(\alpha)}}\ell^{(\alpha)}_m}\chi^* u + R_3,$$

$$T_{\frac{\partial F}{\partial u^{(\alpha)}}} T_{\ell^{(\alpha)}_{m-1}}\chi^* u = T_{\frac{\partial F}{\partial u^{(\alpha)}}\ell^{(\alpha)}_{m-1}}\chi^* u + R_4$$

with $R_3, R_4 \in H^{s+1,k+1}$. Substituting all these into Eq. (4.33) we obtain the required Eq. (4.31). □

According to the above treatment the problem is reduced to the propagation of conormal singularities with respect to the surface $x_1 = 0$ for the solution of Eq. (4.31). Then we can use the method in the proof of Theorem 3.7. Denote by $V_1, V_2, \cdots, V_n$ the operators $x_1\partial_{x_1}, \partial_{x_2}, \cdots, \partial_{x_n}$, then T_{p^*} can be written as the linear combination of the identity I and $V_i, \cdots, V_n$.

Lemma 4.6. *Under the assumptions of Lemma 4.5*

$$T_{p^*} = \sum_{j=1}^{n} T_{B_j} V_j + T_A + Q, \tag{4.34}$$

where T_{B_j} and T_A are paradifferential operators with symbols being polynomials of degree $m-1$, which belong to $C^{s-(n+5)/2,k+1}$ and $C^{s-(n+7)/2,k+1}$ for the variable x respectively, the reminder Q is a continuous map from $H^{s'+m,k'}$ to $H^{s'+s-(n+5)/2,k'}$ with $k' \le k+1$ being a nonnegative integer and s' being any real number.

Proof. Since the symbol p_m^* vanishes on $x_1 = 0$, then it can be written as

$$p_m^* = ax_1\xi_1^m + \sum_{k<m,k+|\alpha|+h=m} b_{k,\alpha,h}\xi_1^k\xi'^{\alpha}\xi_n^h. \tag{4.35}$$

Write $T_{p_m^*}$ as

$$\begin{aligned}T_{p_m^*} &= T_{ax_1}D_1^m + \sum T_{b_{k,\alpha,h}}D_1 D_{x'}^{\alpha}D_n^h \\ &= T_a(x_1 D_1^m) + \sum T_{b_{k,\alpha,h}}D_1 D_{x'}^{\alpha}D_n^h + (T_{ax_1} - T_a x_1)D_1^m,\end{aligned}$$

and incorporate $T_{p^*-p_m^*}$ into T_A and Q, then due to $a \in C^{s-(n+3)/2,k} \subset C^{s-(n+5)/2,k+1}$ we know $T_{ax_1} - T_a x_1$ is a continuous map from $H^{s'+m,k'}$ to $H^{s'+s-(n+5)/2,k'}$. Moreover, notice that $b_{k,\alpha,h}$ is also a $C^{s-(n+3)/2,k}$ function like a, we obtain Eq. (4.34). □

Lemma 4.7. *Assume that $B_j (j = 1, \cdots, n), A$ are paradifferential operators introduced in Lemma 4.6 with homogeneous symbols of degree $m-1$, which are $C^{\rho+1,k'}$ and $C^{\rho,k'}$ $(\rho > 0)$ functions of x respectively. Denote $P = \sum_{j=1}^{n} T_{B_j}V_j + T_A$, then for any integer k and multi-index I, satisfying $1 \le |I| \le k \le k'$, the following relation holds*

$$[V^I, P] = \sum_{|J|\le|I|} T_{B_{J,j}}V_j V^J + R_I. \tag{4.36}$$

Proof. Let $k = 1$, denote $A = \sum a_\alpha \partial_x^\alpha$ and $T_A = \sum T_{a_\alpha}\partial_x^\alpha$, then

$$\begin{aligned}[V_I, T_A] &= \sum [V_I, T_{a_\alpha}]\partial_x^\alpha + \sum V_{a_\alpha}[V_I, \partial_x^\alpha] \\ &= \sum (T_{V_I a_\alpha} + R_{I,\alpha})\partial_x^\alpha + \sum T_{a_\alpha}H_{I,\alpha},\end{aligned}$$

where $R_{I,\alpha}$ is a continuous map from H^{s+1,k_1} to $H^{s+1+\rho,k_1}$ as $k_1 \le k'$, $H_{I,\alpha}$ is a differential operator of order $m-1$, then $[V_I, T_A]$ can be incorporated into R_I in the right-hand side of Eq. (4.36). On the other hand,

$$\begin{aligned}[V_I, T_{B_j}V_j] &= [V_I, T_{B_j}]V_j \\ &= \sum_{|\alpha|=m-1} (T_{V_I, b_{j,\alpha}} + R_{I,j,\alpha})\partial_x^\alpha V_j + \sum T_{b_{j,\alpha}}H_{I,j,\alpha}V_j \\ &= \sum_{|\alpha|=m-1} (T_{V_I, b_{j,\alpha}}\partial_x^\alpha + T_{b_{j,\alpha}}H_{I,j,\alpha})V_j + \sum R_{I,j,\alpha}\partial_x^\alpha V_j. \end{aligned} \tag{4.37}$$

Since the coefficients of B_j are $C^{\rho+1,k'}$ functions, then for $k_1 \le k'-1$, $R_{I,j,\alpha}$ is a continuous map from H^{s+1,k_1-1} to H^{s+1,k_1}, then the second term in the right-hand side of Eq. (4.37) is a continuous map from H^{s+m,k_1} to H^{s+1,k_1}. Meanwhile, the coefficients of V_j in the first term of the right-hand side of Eq. (4.37) is a paradifferential operator with symbol

$$\sum((V_I b_{j,\alpha})\xi^\alpha + b_{j,\alpha} h_{I,j,\alpha}),$$

which belongs to $C^{\rho+1,k'+1}$ with respect to x. Hence Eq. (4.36) hold as $k = 1$.

In the case $k > 1$, the proof can be proceeded by induction. The process of the proof is similar to that for Theorem 3.7. By carefully checking the regularity of all coefficients in the related equalities we found the conclusion of the lemma can be derived from the theory of paradifferential operators. The details are omitted here. □

Proof of Theorem 4.5. First we indicate that $\chi^* u \in H^{s+m,k+1}$ under the assumptions of Lemma 4.5. Indeed, for any multi-index I with $|I| = k+1$, by acting V^I on the both sides of Eq. (4.31) we have

$$T_{p^*} V^I \chi^* u + [V^I, T_{p^*}]\chi^* u = V^I R. \tag{4.38}$$

Applying Lemma 4.7 and taking $k' = k+1$ and $\rho = s - n/2 - 7/2$ we have

$$T_{p^*} V^I \chi^* u = V^I R - \sum_{|J| \le k-1} T_{B_J,j} V_j V^J \chi^* u - R^I \chi^* u. \tag{4.39}$$

From the lemma we know $R \in H^{s+1,k+1}$, then $V^I R \in H^{s+1}$. Moreover, according to Lemma 4.7 we know R^I is a continuous map from $H^{s+m,k}$ to $H^{s+1,0}$, then $R^I \chi^* u \in H^{s+1}$, and $\sum_{|J| \le k-1} T_{B_{J,j}} V_j V^J \chi^* u$ also belongs to H^{s+1}. Hence the right-hand side of Eq. (4.39) is an H^{s+1} function. Therefore, regarding Eq. (4.39) as a system of paradifferential equations for $V^I \chi^* u$ with $|I| \le k+1$, each equation in the system has the same principal part T_{p^*}. Therefore, by using Theorem 4.2 on propagation of singularities for paradifferential equations we can derive from $V^I \chi^* u \in H^{s+m}$ in $x_n < 0$ that $V^I \chi^* u \in H^{s+m}$ in $x_n > 0$ is also true. This means $\chi^* u \in H^{s+m,k+1}$.

Based on the above conclusion and Lemma 4.4 we can alternatively improve the regularity of ϕ and the tangential regularity of $\chi^* u$ on $x_1 = 0$. Hence we have $\phi \in H^\infty$ and $\chi^* u \in H^{s+m,\infty}(x_1 = 0)$. Hence S is a C^∞ smooth surface, and the space $H^{s+m,\infty}(S)$ of conormal distributions is well defined. Furthermore, by using the basic theorem of paracomposition we have (see Theorem A.26 in Appendix)

$$u \circ \chi = \chi^* u + T_{u' \circ \chi} \chi + R, \tag{4.40}$$

where $R \in H^\infty$ as $\chi \in H^\infty$. Hence $u \in H^{s+m,\infty}(S)$, and the conclusion of Theorem 4.5 is obtained. $\square$

The problem on interaction of conormal singularities of solutions to fully nonlinear equations is more complicated, because it is generally hard to prove the characteristic surfaces bearing singularities are C^∞. Then one can only inductively define tangential vector fields and the corresponding space of conormal distributions. For the case of interaction of two characteristic surfaces bearing conormal singularities we refer readers to [4].

Chapter 5

Propagation of strong singularities for nonlinear equations

In the previous chapters, we discuss the propagation of singularities of solutions to nonlinear partial differential equations, in order to let all nonlinear functions be well defined we always require that the solutions are H^s functions at least. When $\partial^k u$ appears in nonlinear functions as arguments the requirement will be $u \in H^{s+k}$. Hence the singularities of solutions under discussion are rather weak. However, in practice stronger singularities are also the object of our study, and are even more important in many cases. For instance, shocks, rarefaction waves and contact discontinuities in gas dynamics, are main nonlinear waves, which people are concerned with. In material sciences the crack of material gives the strong singularity of displacement, which is often the focus of the related discussions. Therefore, in this and next chapter we are going to pay more attention on analysis of strong singularities.

Different from the study of propagation of weak singularities, in the discussion of propagation of strong singularities we generally have to consider the existence of solution and the description of singularities together. Therefore, the description of the singularities of solutions is often related to the construction of solutions with specific singularity structure.

Generally we distinguish the singularity structure in multidimensional case as fan-shaped structure and flower-shaped structure. The fan-shaped structure is an one-dimensional like structure with multidimensional perturbation, while the flower-shaped structure is essentially multidimensional. In Sections 5.1 and 5.2 we discuss two sample problems for semilinear hyperbolic equations corresponding to these two different structure. In Sections 5.3 and 5.4 we turn to discuss the cases for quasilinear equations, which are closely related to gas dynamics and other applied sciences. Due to the rapid development and greatly plentiful results in this area we only give a survey on the topics and introduce corresponding references to readers there.

5.1 Solutions with fan-shaped singularity structure of semi-linear equations

In this section we are going to study the solutions with fan-shaped singularity structure. The character of such a structure is one-dimensional like structure with multidimensional perturbation.

Consider a Cauchy problem of a hyperbolic equation with initial data having discontinuity on a submanifold σ of codimension 1 on the initial plane, the solution of the Cauchy problem may also have discontinuity on some surfaces issuing from σ. For semilinear equations these surfaces are characteristics, while for quasilinear equations or fully nonlinear equations singularities can propagate along more general surfaces. In any case, all the surfaces carrying singularities issuing from the given submanifold σ on the initial plane like a fan with ridge σ, so that is called **fan-shaped structure**. Since the regularity of the solutions on different directions of the surfaces bearing discontinuity are quite different, it is generally not suitable to describe the singularities of those solutions by using wave front sets. Conversely, the conormal distributions or piecewise smooth functions may give good description of singularity structure.

Consider the following strictly hyperbolic equations with respect to t:

$$Pu \equiv \left(A_1 \partial_t + A_2 \partial_x + \sum_{j=3}^{n} A_j \partial_{y_j} \right) u = F(t, x, y, u), \tag{5.1}$$

where $A_1, \cdots, A_n$ are 3×3 symmetric matrix, F and all coefficients are C^∞ smooth. Assume that the initial data are given on $t = 0$, which has discontinuity of first class on a submanifold $\sigma \subset \{t = 0\}$ of dimention $n - 2$. Denote $\Sigma_1, \Sigma_2, \Sigma_3$ are characteristic surfaces issuing from σ, we have the following theorem:

Theorem 5.1. *There is a unique piecewise smooth solution u of the Cauchy problem, which only has discontinuity of first class on Σ_i $(i = 1, 2, 3)$.*

The proof will be carried out by using several lemmas. Assume that Σ_1 (Σ_2 resp.) is the right (left resp.) characteristic surfaces issuing from σ, while Σ_3 is the characteristic surface in the middle. According to the property of finite propagation speed for hyperbolic equations we can determine the solution of the hyperbolic equation (5.1) in the right side of Σ_1 and the left side of Σ_2. Therefore, the problem is reduced to a Goursat problem. By using the equation and the initial data, the jump of the solution u and the derivatives of u on σ can be determined, so that the boundary data on Σ_1, Σ_2 of u can be determined. Finally, by constructing a series of estimates and discretizing the derivatives with respect to y we can derive the existence of solution with assigned singularity structure.

First let us straighten the characteristic surfaces. Assume that the equation of Σ_i is $x = \psi_i(t, y)$ $(i = 1, 2, 3)$, satisfying

$$\begin{aligned} &\psi_2(t, y) < \psi_3(t, y) < \psi_1(t, y), \\ &\partial_t\psi_2(t, y) < \partial_t\psi_3(t, y) < \partial_t\psi_1(t, y). \end{aligned} \tag{5.2}$$

Lemma 5.1. *There is a C^∞ reversible transformation, which transforms Σ_1,Σ_2,Σ_3 into $x - t = 0, x + t = 0, x = 0$ respectively.*

Proof. Since in the discussion of this lemma the coordinates y are parameters and are always unchanged, we will omit y for the notational simplicity. First, the transformation

$$\begin{cases} x' = x + \dfrac{1}{2}(\psi_1(t, y) + \psi_2(t, y)), \\ t' = \dfrac{1}{2}(\psi_1(t, y) - \psi_2(t, y)), \end{cases} \tag{5.3}$$

can transform Σ_1, Σ_2 into $x' - t' = 0, x' + t' = 0$. Without loss of generality we assume that the equations of Σ_1, Σ_2 are $x' - t' = 0$ and $x' + t' = 0$. It means that we may assume $\psi_1(t, y) \equiv t, \psi_2(t, y) \equiv -t$. Moreover, Eq. (5.2) implies $|\partial_t\psi_3(t, y)| < 1$.

Let

$$x' = \frac{1}{\sqrt{2}}(t + x), \quad t' = \frac{1}{\sqrt{2}}(t - x).$$

$\Sigma_1, \Sigma_2, \Sigma_3$ are transformed to $t' = 0$, $x' = 0$ and

$$F(t', x', y') \equiv \frac{1}{\sqrt{2}}(x' - t') - \psi_3\left(\frac{1}{\sqrt{2}}(t' + x'), y\right) = 0. \tag{5.4}$$

Since $\dfrac{\partial F}{\partial x'} = \dfrac{1}{\sqrt{2}}(1 - \partial_t\psi_3) > 0$, then the equation of Σ_3 can be written as $x' = g(t', y)$. Making another transformation

$$t'' = t' \quad x'' = \frac{t'}{g(t', y)}x',$$

then the three characteristic surfaces are transformed to $t''=0, x''=0$ and $t''=x''$. Finally, the transformation

$$\bar{t}=\frac{1}{\sqrt{2}}(t''+x''), \quad \bar{x}=\frac{1}{\sqrt{2}}(-t''+x'')$$

let the three characteristic surface become $\bar{t}=\bar{x}, \bar{t}=-\bar{x}$ and $\bar{x}=0$. □

Remark. It is not difficult to combine all above transformations into one transformation, but we prefer to write all details to show the process of constructing the transformation.

Again denote the variables by t, x instead of $\bar{t}, \bar{x}$, by virtue of above lemma Eq. (5.1) has new form as

$$X_\ell u_\ell + \sum_{i=1}^{3}\sum_{j=1}^{n} b_{\ell ij}\partial_{y_j} u_i = f_\ell \quad (\ell=1,2,3), \tag{5.5}$$

where $X_\ell=\partial_t+\lambda_\ell\partial_x$. In addition, λ_ℓ satisfy: $\lambda_1=1$ on $\Sigma_1(x=t)$, $\lambda_2=-1$ on $\Sigma_2(x=-t)$ and $\lambda_3=0$ on $\Sigma_3(x=0)$. The solution in the region right to Σ_1 and the region lift to Σ_2 can be regarded as known.

Lemma 5.2. *The jump of u and its all derivatives on Σ_ℓ can be determined by using its data on $t=0$ and Eq. (5.5).*

Proof. For any function w defined in $t\geq 0$, we denote by $[w]_{\ell,0}$ the jump of w on Σ_ℓ at $t=0$. For any function w^0 define on $t=0$ we denote by $[w^0]_0$ the jump of w^0 at $x=0$. Next we indicate that all jumps $[X^\alpha\partial_y^\beta u_\ell]_{\ell',0}$ can be determined by induction.

First, on $\gamma:\ t=x=0$ the differential operator $\partial/\partial y$ is tangential to γ and all characteristic surfaces Σ_ℓ, then the differentiation with respect to y will not cause any difficulty. Hence in the following discussion we can only consider the case $\beta=0$.

Consider the case $\alpha=0$. Notice that X_ℓ is transversal to $\Sigma_{\ell'}$, as $\ell\neq\ell'$. Then Eq. (5.5) gives $[u_\ell]_{\ell',0}=0$, as $\ell\neq\ell'$. Furthermore,

$$[u_\ell]_{\ell,0}=[u_\ell^0]_0. \tag{5.6}$$

Now assume that all $[X^\alpha u_\ell]_{\ell',0}$ with $|\alpha|\leq m$ are known, we are going to determine $[X^\alpha u_\ell]_{\ell',0}$ with $|\alpha|\leq m+1$. To this end we write $[X^\alpha u_\ell]_{\ell',0}$ as

$$[X^{\alpha_1}X_i u_\ell]_{\ell',0}, \tag{5.7}$$

where $|\alpha_1|=m$. Obviously, in the case $i=\ell$, by using Eq. (5.5) the expression (5.7) can be reduced to the case when only m vector fields in

X^{α_1} acting on u. Hence it can be determined by the initial data according to the hypothesis of induction.

In the case $i \neq \ell$ and $\ell \neq \ell'$, we notice that any vector field among three vector fields X_1, X_2, X_3 is a linear combination of other two, then writing any factor in $X^{\alpha_1} X_i$ by the linear combination of X_ℓ and $X_{\ell'}$ we obtain

$$[X^{\alpha_1} X_i u_\ell]_{\ell',0} = [a X_{\ell'}^{m+1} u_\ell]_{\ell',0} + [X^{\alpha_2} X_\ell u_\ell]_{\ell',0} + \text{known},$$

where "known" means the terms, which can be determined by initial data. Since $X_{\ell'}$ is tangential to $\Sigma_{\ell'}$, then all terms in the right-hand side of above equality are known in fact.

The remaining case is $i \neq \ell$ and $\ell = \ell'$. For instance, for $\ell = 1$ we have

$$\begin{aligned}[X^{\alpha_1} X_i u_1]_{1,0} &= (X^{\alpha_1} X_i u_1)_{t=0,x=+0} - (X^{\alpha_1} X_i u_1)_{t=0,x=-0} \\ &\quad - [X^{\alpha_1} X_i u_1]_{2,0} - [X^{\alpha_1} X_i u_1]_{3,0}.\end{aligned}$$

The first term in the right-hand side can be determined by the initial data, while the second and the third terms are known according to above discussion for other two cases. □

Lemma 5.3. *All derivatives of u along Σ_1, Σ_2 can be determined by using the initial data and Eq. (5.5).*

Proof. By using the jumps of u and its all derivatives on characteristic surfaces at the origin we can determine the jumps of them on whole characteristic surfaces. Next we will also prove them by induction. As the proof of Lemma 5.2 we do not need to consider the derivatives with respect to y.

First, Eq. (5.1) shows that the derivatives of u_1, u_3 in he direction transversal to Σ_2 are bounded, then

$$[u_1]_{\Sigma_2} = [u_3]_{\Sigma_2} = 0.$$

By the same reason

$$[u_2]_{\Sigma_1} = [u_3]_{\Sigma_1} = 0.$$

Consider the equations in Eq. (5.5) corresponding to $\ell = 1$ on both sides of Σ_1, making substraction gives

$$X_1[u_1]_{\Sigma_1} + \sum b_{11j} \partial_{y_j} [u_1]_{\Sigma_1} = f_1(u|_{\Sigma_1^+}) - f_1(u|_{\Sigma_1^-}), \tag{5.8}$$

where f_1 is a known function. Obviously, $u|_{\Sigma_1^+}$ is known, and $u|_{\Sigma_1^-} = u|_{\Sigma_1^+} - [u]_{\Sigma_1}$, Hence Eq. (5.8) can be regarded as a differential equation of $[u_1]_{\Sigma_1}$. Since the initial data of $[u_1]_{\Sigma_1}$ at σ is known, then the value of $[u_1]_{\Sigma_1}$ on Σ_1 can be determined by solving the differential equation. Similarly, $[u_2]_{\Sigma_2}$ can be obtained.

Now if we know all $[X^\alpha u_\ell]_{\Sigma_{1,2}}$ with $\ell = 1,2,3$ and $|\alpha| \le m$, then for $\ell = 1,2$ and $\ell' = 1,2,3$

$$[X_\ell X^\alpha u_{\ell'}]_{\Sigma_\ell} = X_\ell [X^\alpha u_{\ell'}]_{\Sigma_\ell}$$

are also known. From Eq. (5.5) itself we know that for $\ell = 1,2$

$$[X_3 X^\alpha u_3]_{\Sigma_\ell} = \left[X^\alpha \left(\sum b_{3ij}\partial_{y_j} u_i\right)\right]_{\Sigma_\ell} + [[X_3, X^\alpha]u_3]_{\Sigma_\ell},$$

which can be regarded as known by using the hypothesis of the induction.

Since the jumps $[X_\ell X^\alpha u_i]_{\Sigma'_\ell}$ with ℓ, ℓ', i being different from each other can always be expressed as linear combination of such jumps with $\ell = i$ or $\ell = \ell'$. Hence we only have to estimate $[X_1 X^\alpha u_2]_{\Sigma_2}$ and $[X_2 X^\alpha u_1]_{\Sigma_1}$. On both sides of Σ_1 by acting $X_2 X^\alpha$ on the first equation of Eq. (5.5), we have

$$X_1[X_2 X^\alpha u_1]_{\Sigma_1} + \sum b'_j \partial_{x_j}[X_2 X^\alpha u_1]_{\Sigma_1} = \text{ known} \tag{5.9}$$

by direct computations. Regarding Eq. (5.9) as a differential equation for $[X_2 X^\alpha u_1]_{\Sigma_1}$, whose initial data on σ has been determined according to Lemma 5.2, we obtain the quantity on the surface Σ_1. Similarly, $[X_1 X^\alpha u_2]_{\Sigma_2}$ can also be determined. Hence we obtain all $[X^\alpha u_\ell]_{\Sigma_{1,2}}$ for $\ell = 1,2,3$ and any α satisfying $|\alpha| = m + 1$. □

Next we are going to establish the energy estimates for the solution to the Goursat problem of Eq. (5.5), which can lead us to the existence of solution directly. To this end we introduce some Sobolev spaces, which are powerful in treating the characteristic boundary value problems of hyperbolic equations. Let $\partial^{s,s'}u$ denote $\partial^s \partial_\tau^{s'} u$, where ∂_τ means the tangential derivatives and ∂ means the regular derivatives without any restriction. Define

$$\|u\|_{s,s'} = \left(\sum_{r \le s, r' \le s'} \|\partial^{r,r'} u\|^2 \right)^{1/2},$$

$$\|u\|_{B^s} = \sum_{2r \le s} \|u\|_{r,s-2r},$$

$$\|u\|_{\hat{B}^s} = \sum_{2r \le s} \|u\|_{r+1,s-2r-1}.$$

The above norm describes such a property of functions: the increase of regularity of u of order 1 in non-tangential direction should be compensated by the loss of its regularity of order 2 in tangential direction. For the Sobolev spaces B^s and $\hat{B}^s$ the following proposition holds.

Lemma 5.4. *For $s > n$, the spaces B^s and $\hat{B}^s$ are invariant under nonlinear composition.*

Proof. According to the Sobolev embedding theorem $H^{s,s'} \subset L^\infty$ if $s \geq 1$, $s + s' > n/2$. Hence for $s > n$ we have $B^s \subset L^\infty$. Besides, $H^{0,s} \cap L^\infty$ is invariant under nonlinear composition for $s > n/2$. It is easy to verify that $H^{r,s-2r}$ is invariant under nonlinear composition for $r > 1/2$, $r+(s-2r) > n/2$. Hence B^s and $\hat{B}^{s'}$ are invariant under nonlinear composition. □

Lemma 5.5. *Let Ω_τ be the domain $|x| < t, t \leq \tau$, $\Sigma_{i\tau}$ be the surface $\Sigma_i \cap \{t \leq \tau\}$. If u is a solution of Eq. (5.5), then for $s > n$*

$$\|u_3\|_{B^s(\Omega_\tau)} + \|u_{1,2}\|_{\hat{B}^s(\Omega_\tau)} \leq C \sum_{\ell \leq s} \|\partial^\ell u\|_{H^{s-\ell}(\Sigma_{1\tau} \cup \Sigma_{2\tau})}. \tag{5.10}$$

Proof. In the domain Ω we introduce the transformation $x \to -x$, which can fold the domain Ω to Ω^+, bounded by Σ_1 and Σ_3. Now let

$$u_{i+3}(t,x,y) = u_i(t,-x,y) \quad (i = 1,2,3),$$

we obtain a differential system of $U = (u_1, \cdots, u_6)^T$

$$\bar{A}U + \sum \bar{B}_j \partial_{y_j} U = F, \tag{5.11}$$

where

$$\bar{A} = \mathrm{diag}(X_1, X_2, X_3, X_1', X_2', X_3'),$$

$$X_i' = \partial_t - \lambda_i(t,-x,y)\partial_x \quad (i = 1,2,3),$$

the expressions of $\bar{B}_j, F$ can also be obtained form Eq. (5.5) by direct computations. As for the boundary conditions, the value of all components of U and its derivatives on Σ_1 can be determined by using Lemma 5.3, and the conditions on Σ_3 are

$$u_4 - u_1 = 0, \quad u_5 - u_2 = 0. \tag{5.12}$$

The method to derive the energy estimates is classical. Denote the upper bound of Ω_τ by π_τ, by multiplying U on Eq. (5.11) and integrating on Ω_τ^+ we have

$$\begin{aligned}\int_{\pi_\tau} U^2 dxdy \leq & \int_{\Omega_\tau^+} (U^2 + F^2) d\tau dxdy + \frac{1}{\sqrt{2}} \int_{\Sigma_{1\tau}} U^2 dS \\ & + \int_{\Sigma_{3\tau}} (u_2^2 - u_1^2 + u_4^2 - u_5^2) dS.\end{aligned} \tag{5.13}$$

The last integral vanishes due to Eq. (5.12). Then we obtain

$$\int_{\pi_\tau} U^2 dxdy \leq C \left(\int_{\Omega_\tau^+} F^2 d\tau dxdy + \int_{\Sigma_{1\tau}} U^2 dS \right) \tag{5.14}$$

by using Gronwall inequality.

Since $\partial_t, x\partial_x, \partial_{y_j}$ are tangential operators to Σ_3, then we can act these operators on the boundary conditions shown in Eq. (5.12). By using the same method as above we obtain

$$\int_{\pi_\tau} (\partial_t^\alpha (x\partial_x)^\beta \partial_y^\gamma U)^2 dxdy \le C\left(\int_{\Omega_\tau^+} F_1^2 d\tau dxdy + \int_{\sigma_{1\tau}} (\partial_t^\alpha (x\partial_x)^\beta \partial_y^\gamma U)^2 dS\right), \tag{5.15}$$

where F_1 satisfies

$$\int_{\Omega_\tau^+} F_1^2 dtdxdy \le C(\|F\|_{0,s}^2 + \|U\|_{0,s}^2).$$

Hence we have

$$\|\partial^{0,s} U\|_{\pi_\tau}^2 \le C(\|F\|_{H^{0,s}(\Omega_\tau^+)}^2 + \|U\|_{H^{0,s}(\Sigma_\tau)}^2).$$

Since X_1, X_1', X_2, X_2' are transversal to Σ_3, then

$$\|\partial^{1,s-1} u_{1,2,4,5}\|_{\pi_\tau}^2 + \|\partial^{0,s} u_{3,6}\|_{\pi_\tau}^2 \le C(\|F\|_{H^{0,s}(\Omega_\tau^+)}^2 + \|U\|_{H^{0,s}(\Sigma_\tau)}^2). \tag{5.16}$$

Next we are going to prove

$$\|\partial^{r+1,s-2r-1} u_{1,2,4,5}\|_{\pi_\tau}^2 + \|\partial^{r,s-2r} u_{3,6}\|_{\pi_\tau}^2 \le C(\|F\|_{B^s(\Omega_\tau^+)}^2 + \|\partial^s U\|_{L^2(\Sigma_\tau)}^2) \tag{5.17}$$

for $r \ge 1$ by induction. Obviously, the inequality is valid for $r = 0$ by virtue of Eq. (5.16). Now we assume that Eq. (5.17) is valid for r satisfying $s/2 - 1 \ge r \ge 0$ and prove its validity for $r + 1$. Indeed, for any multi-index (α, β, γ) satisfying $\alpha + \beta + |\gamma| \le s - 2r - 2$, by acting the operator $L = X_2^{r+1} \partial_t^\alpha (x\partial_x)^\beta \partial_y^\gamma$ on the third equation in Eq. (5.15) we obtain

$$\begin{aligned} X_3 L u_3 + [L, X_3] u_3 + \sum (b_{33j} \partial_{y_j} L u_3 + [L, b_{33j} \partial_{y_j}] u_3 \\ + L b_{31j} \partial_{y_j} u_1 + L b_{32j} \partial_{y_j} u_2) = L f_3, \end{aligned}$$

which can also be written as

$$X_3 L u_3 + \sum b_{33j} \partial_{y_j} L u_3 + a L u_3 = g, \tag{5.18}$$

where

$$\|g\|_{L^2(\Omega_\tau^+)} \le C(\|f_3\|_{\gamma+1,s-2r-2} + \|u_{1,2}\|_{r+1,s-2r-1}).$$

Hence by using the same method in deriving (5.14) we can obtain

$$\|L u_3\|_{\pi_\tau}^2 \le C(\|f_3\|_{\gamma+1,s-2r-2}^2 + \|u_{1,2}\|_{r+1,s-2r-1}^2 + \|\partial^{r+1,s-2r-1} u_3\|_{\Sigma_{1,\tau}}^2),$$

which implies

$$\|\partial^{r+1,s-2r-2} u_{3,6}\|_{\pi_\tau}^2 \le C(\|F\|_{B^s(\Omega_\tau^+)}^2 + \|\partial^s U\|_{\Sigma_{1\tau}}^2). \tag{5.19}$$

Similarly, by using the first and second equations in Eq. (5.5) we can control $\|\partial^{r+2,s-2r-3}u_{1,2,4,5}\|^2_{\pi_\tau}$ by the right-hand side of (5.17). This means that (5.17) with r replaced by $r+1$ still holds. Therefore, (5.17) holds for any r.

Finally, by using Lemma 5.4 the term $\|F\|_{B^s(\Omega^+_\tau)}$ is controlled by $C\|U\|_{B^s(\Omega^+_\tau)}$. (5.10) is thus established. □

The proof of Theorem 5.1

Rewrite Eq. (5.5) as

$$\begin{cases} \tilde{X}_1 u_1 + \sum (A_j)_{12}\partial_{y_j} u_2 + \sum (A_j)_{13}\partial_{y_j} u_3 = f_1(u), \\ \tilde{X}_2 u_2 + \sum (A_j)_{21}\partial_{y_j} u_1 + \sum (A_j)_{23}\partial_{y_j} u_3 = f_2(u), \\ \tilde{X}_3 u_3 + \sum (A_j)_{31}\partial_{y_j} u_2 + \sum (A_j)_{32}\partial_{y_j} u_3 = f_3(u), \end{cases} \tag{5.20}$$

where $\tilde{X}_i = X_i + \sum (A_j)_{ii}\partial_{y_j}$, and $\sum$ means the summation for $3 \le j \le n$. Replacing the derivatives with respect to y_j by difference quotient

$$L_h^{(ik)} u_k(t,x,y) = \sum (A_j)_{ik}(t,x,y)\frac{1}{2h}(u_k(t,x,y+he_j) - u_k(t,x,y-he_j)),$$

Eq. (5.5) becomes

$$\begin{cases} \tilde{X}_1 u_1 + L_h^{(12)} u_2 + L_h^{(13)} u_3 = f_1(u), \\ \tilde{X}_2 u_2 + L_h^{(21)} u_1 + L_h^{(23)} u_3 = f_2(u), \\ \tilde{X}_3 u_3 + L_h^{(31)} u_1 + L^{(32)} u_2 = f_3(u). \end{cases} \tag{5.21}$$

Besides, we can also make difference quotient for the boundary condition Eq. (5.12) to establish a Goursat problem of Eq. (5.21). The existence of the Goursat problem can be established by using characteristic method. Since $L_h^{ik} + (L_h^{ki})^*$ is uniformly bounded with respect to h, then the solutions to the Goursat problem has a uniform energy estimate with respect to h. Therefore, we can choose a subsequence of solutions, which is convergent as $h \to 0$. The limit satisfies the estimate Eq. (5.10). □

Theorem 5.2. *Under the assumptions of Theorem 5.1, if the initial data is smooth at the point $P \in \sigma$, then the solution is smooth on the bicharacteristics of Σ_i passing through P.*

Proof. Let us consider the case $i = 1$. By using a suitable transformation of variables and unknown functions, Eq. (5.1) can be written as

$$\sum_{j=1}^{n} B_j \frac{\partial V}{\partial z_j} = G(V), \tag{5.22}$$

where $z_1 = 0$ is the image of Σ_1, $B_1 = diag(0, B_1')$ with B_1' being an $(m-1)\times(m-1)$ non-singular matrix. Denote $(V_2, \cdots, V_m)$ by V', then $V = (V_1, V')$ and Eq. (5.22) can be written as

$$\begin{cases} \ell_1 V_1 = G_1 - \displaystyle\sum_{k=2}^{m} \ell_k V_k, \\ \displaystyle\sum_{j=1}^{n} B_j' \frac{\partial V'}{\partial z_j} = G' - h' V_1, \end{cases} \tag{5.23}$$

where $\ell_k (k \geq 2), h'$ are differential operators defined on $z_1 = 0$, $\ell_1 = \displaystyle\sum_{j=2}^{n} (B_j)_{11} \frac{\partial}{\partial z_j}$.

The symbol matrix of Eq. (5.22) is $p(z,\zeta) = \det|\sum B_j\zeta_j|$. Since the equation of $N^*(\Sigma_1)$ is $z_1 = 0, \zeta_2 = \cdots = \zeta_n = 0$, then by using the special form of B_1 we have

$$\frac{\partial p}{\partial \zeta_1} = 0, \quad \frac{\partial p}{\partial \zeta_j} = (B_j)_{11} \det|B_1'|.$$

In accordance, the bicharacteristics L_1 issuing from P with initial direction $(1, 0, \cdots, 0)$ is the integral curve of the vector field $\ell_1 = \displaystyle\sum_{j=2}^{n} (B_j)_{11} \frac{\partial}{\partial z_j}$.

By using Eq. (5.23) and the C^∞ regularity of V at $P \in \sigma$, we obtain the C^∞ regularity of V along L_1. Indeed, V is the solution of Eq. (5.23), and B_1' is a non-singular matrix, then the normal derivative of V' can be expressed by using V and its tangential derivatives. The fact means V' is C^0 on $x_1 = 0$. Furthermore, from the first equation in Eq. (5.23) and the regularity of V_1 at P we know V_1 is C^0 along L_1. Then we can improve the regularity of V continuously by induction. That is, assume that V is C^k on L^1, then V' is C^{k+1} on L^1 by using the second equation. Furthermore, from the first equation in Eq. (5.23) and the regularity of V_1 at P we know V_1 is C^{k+1} along L_1. Therefore, we know $V = (V_1, V')$ in C^∞ along L_1 by induction. □

Theorems 5.1 and 5.2 indicate the following theorem on interaction of singularities.

Theorem 5.3. *Assume that u is a solution of Eq. (5.1), Σ_1, Σ_2 are two characteristic surfaces separated from each other in $t < 0$ and intersect at a space-like curve γ in $t > 0$, then the following conclusions hold.*

(1) If u has discontinuity of first class only on Σ_1 as $t < 0$, then the discontinuity will propagate to $t > 0$ along Σ_1.

(2) If u has discontinuity of first class only on Σ_1 and Σ_2 as $t < 0$, then besides the propagation of discontinuity of u along Σ_1 and Σ_2, the singularity of u may also appear on the characteristic surface Σ_3 issuing from γ.

In fact, since γ is a space-like curve, then we can make a space-like surface Π passing γ. By flattening Π to a plane we obtain a Cauchy problem of hyperbolic equation with discontinuous initial data, which has been discussed in Theorem 5.1, so that the conclusion of Theorem 5.3 can be obtained from Theorem 5.1 directly.

The discussion in this section indicates that in the framework of piecewise smooth functions when two characteristic surfaces Σ_1, Σ_2 bearing strong singularity intersect transversally, besides the propagation of singularity along these two surfaces, some extra singularity may appear due to interaction, which can propagate along a new characteristic surface Σ_3. Certainly, such an interaction phenomenon does not occur for 2×2 system. As for $n \times n$ system with $n > 3$, we could not flatten all characteristics by a coordinate transformation. In this case, conormal distributions can be applied to describe singularity of solutions. Here we introduce a result in [104] as follows.

Consider a Cauchy problem

$$\begin{cases} \partial_t u + \sum_{i=1}^{n} A_i(t,x)\partial_{x_i} u = F(t,x,u), \\ u\big|_{t=0} = g(x), \end{cases} \tag{5.24}$$

where $g(x)$ has singularity on σ with codimension 1. Let Ω be an open set in R^{n+1}_+, which is contained in the determinacy domain of $\Omega \cap \{t = 0\}$. Denote by $\Omega_T \subset R^{n+1}_+$ the domain $\Omega \cap \{t < T\}$, and denote by $H^{0,m}(\Omega_T)$ the conormal distributions as defined in Chapter 3. Then we have

Theorem 5.4. *If $m > (n+5)/2, g \in L^\infty(R^n) \cap H^{0,m}(R^n)$, then there is a unique solution u in Ω_T. Moreover, $u \in L^\infty(\Omega_T) \cap H^{0,m}(\Omega_T)$.*

The theorem indicates the singularity of the initial data of u on σ will propagate along all characteristic surfaces issuing from σ. Readers may refer [104] for the proof of the theorem.

5.2 Solutions with flower-shaped singularity structure of semilinear equations

In this section we are going to study the solutions with flower-shaped singularity structure. In this case singularities of solutions propagate from a point on the initial plane to outside, so that the surfaces (or lines) bearing singularities like a flower with a vertex at the given point. Since such a structure cannot be obtained by just adding a perturbation to an one-dimensional wave structure in multidimensional space, it is also called **essentially multidimensional wave structure**.

Let us discuss a specific Cauchy problem for a semilinear wave equation to understand flower-shaped singularity structure. The problem is

$$\begin{cases} \Box u = f(t,x,y,u), \\ u(0,x,y) = 0, \quad u_t(0,x,y) = \psi(x,y), \end{cases} \tag{5.25}$$

where $\Box = \partial_{tt} - \partial_{xx} - \partial_{yy}$, $f(t,x,y,u)$ is a C^∞ function of their arguments, and may also be written as $f(u)$ for simplicity. Assume that $\psi(x,y)$ takes four different constants in four quadrants, i.e.

$$\psi_i(x,y) = c_i, \quad \text{in } i\text{-th quadrants}. \tag{5.26}$$

Such a problem is also called **two dimensional Riemann problem** for the semilinear wave equation. The name comes from the multidimensional Riemann problem for the system of gas dynamics. Indeed, the multidimensional Riemann problem for the system of gas dynamics is much more complicated, because the corresponding system is a quasilinear system.

Theorem 5.5. *For small $T_0 > 0$ there is a C^0 solution u of Eq. (5.25) in $0 \le t \le T_0, -\infty < x < \infty, -\infty < y < \infty$, which is C^∞ outside the characteristic surfaces $x = \pm t, y = \pm t$ and $t^2 = x^2 + y^2$. u is a conormal distribution with respect to all these surfaces. Moreover, u actually has singularity on these surfaces.*

In order to prove the theorem we first give an outline of the proof. Assume that v is a solution of linear wave equation with the same initial data as shown in Eq. (5.26), E is the inverse of wave operator taking homogeneous initial condition, then the solution u of Eq. (5.26) satisfies

$$u = v + Ef(t,x,y,u). \tag{5.27}$$

If we can find a suitable space B, such that the following facts hold:

(1) $v \in B$,

(2) $Ef(t,x,y,u) \in B$, if $u \in B$,

(3) For small T, the map $u \to Ef(t,x,y,u)$ is a contractive map in some sense.

Then the existence of the solution of Cauchy problem Eq. (5.25) can be established via iteration. Meanwhile, the space B gives the singularity structure of the solution.

Let us first observe, where the singularity of the solution u could appear. The wave front set of the initial data $\psi(x,y)$ is $(x,0;0,1),(0,y;1,0)$ and $(0,0,;\xi,\eta)$ in $R^2_{xy} \times R^2_{\xi\eta}$. The projection of all bicharacteristic strips corresponding those wave front set on the space (t,x,y) is

$$x = \pm t,\ y = \pm t,\ t^2 = x^2 + y^2. \tag{5.28}$$

Hence we will construct the space B, so that any function in B has conormal regularity with respect to the characteristic surfaces in (5.28).

Denote $\Sigma_1 = \{x = t\}$, $\Sigma_2 = \{y = t\}$, $\Sigma_3 = \{x = -t\}$, $\Sigma_4 = \{y = -t\}$, $\Sigma_5 = \{t^2 = x^2 + y^2\}$, $S = \{\Sigma_1, \cdots, \Sigma_5\}$. Denote $\ell_i = \Sigma_i \cap \Sigma_5 (i = 1,2,3,4)$, $\ell_5 = \Sigma_1 \cap \Sigma_2$, $\ell_6 = \Sigma_2 \cap \Sigma_3$, $\ell_7 = \Sigma_3 \cap \Sigma_4$, $\ell_8 = \Sigma_4 \cap \Sigma_1$. Assume that $\{\omega_j\}$ is an open conical covering of R^3_+, satisfying the following conditions.

(1) Each ω_j is contained in a cone with vertex angle less than $\pi/10$,

(2) $\ell_j \subset \omega_j$, $\ell_j \cap \omega_{j'} = \emptyset$, if $j, j' \leq 8$ and $j \neq j'$,

(3) When $j > 8$, ω_j intersects at most one surface in S, while for $j \leq 8$ each ω_j intersect two surfaces in S.

Then in any conical neighborhood ω_j we can find a complete basis of tangential vector fields $M_1, \cdots, M_\ell$ with respect to S. For instance, among all characteristic surfaces Σ_i with $1 \leq i \leq 8$, only Σ_1 and Σ_5 appear in ω_1, then the basis in ω_1 can be chosen as

$$\begin{aligned} M_0 &= t\partial_t + x\partial_x + y\partial_y, \\ M_1 &= t\partial_x + x\partial_t, \\ M_2 &= y(\partial_t + \partial_x) + (t - x)\partial_y. \end{aligned} \tag{5.29}$$

The choice of the complete basis in $\omega_2, \omega_3, \omega_4$ is the same as that in ω_1. For ω_5, the surfaces Σ_1 and Σ_2 appear there, so that the complete basis can be chosen as

$$\begin{aligned} M_0 &= t\partial_t + x\partial_x + y\partial_y, \\ M_3 &= (x - y)\partial_t + (t - y)\partial_x + (x - t)\partial_y, \\ N &= t(\partial_t + \partial_x + \partial_y). \end{aligned} \tag{5.30}$$

The choice of the complete basis in $\omega_6, \omega_7, \omega_8$ is the same as that in ω_5. When $j > 8$, ω_j only intersects one surface in S, then the choice of the complete basis is easier (see Chapter 3). In the sequel we often take ω_1 and ω_5 as two typical neighborhoods to derive related estimates, because for other conical neighborhoods, the situation is similar or even simpler.

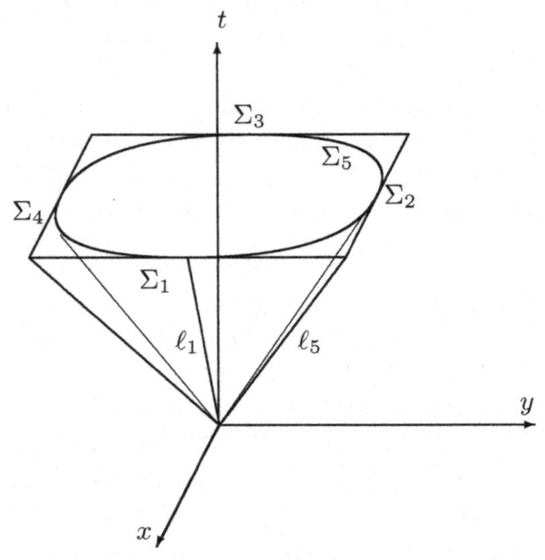

Figure 5.1. Flower-shaped singularity structure for a semilinear equation.

Construct a partition of unity $\sum \chi_i \equiv 1$ subordinated to the open conical covering $\{\omega_j\}$, so that each χ_i is a homogeneous function of degree 0. In each conical neighborhood ω_j, if $\{M_1, \cdots, M_\ell\}$ is the basis of vector fields tangential to S, then we can define $H^{0,k}(\omega_i)$ as

$$\{u;\ M_1^{\alpha_1} M_2^{\alpha_2} \cdots M_\ell^{\alpha_\ell}(\chi_j u) \in L^2(\omega_j \cap C_T)\},$$

where C_T is the domain $(t-T)^2 > x^2 + y^2$. Correspondingly, the space $H^{0,k}$ can be defined as

$$\{u;\ \chi_i u \in H^{0,k}(\omega_i) \text{ for each } i\}.$$

Take $L^\infty \cap H^{0,k}$ as the space B, we have the following lemma.

Lemma 5.6. *The solution v of the linear wave equation satisfying the initial condition Eq. (5.26) belongs to B.*

Proof. The solution v can be expressed by using Poisson formula as

$$v = \frac{1}{2\pi} \int_{r \le t} \frac{\psi(\xi, \eta)}{\sqrt{t^2 - (x-\xi)^2 - (y-\eta)^2}} d\xi d\eta, \tag{5.31}$$

where $r = ((x-\xi)^2 + (y-\eta)^2)^{1/2}$. Substituting the value of $\psi(x,y)$ into Eq. (5.31) gives more detailed expression. Due to the symmetry we only write the expression in $x \geq 0, y \geq 0$.

$$v = \begin{cases} c_1 t, \quad \text{if } \; x > t, y > t, \\ \frac{t}{2}(c_1 + c_2) + \frac{x}{2}(c_1 - c_2), \quad \text{if } \; 0 < x < t, y > t, \\ \frac{t}{2}(c_1 + c_4) + \frac{y}{2}(c_2 - c_4), \quad \text{if } \; 0 < y < t, x > t, \\ \frac{t}{2}(c_2 + c_4) + \frac{x}{2}(c_1 - c_2) + \frac{y}{2}(c_1 - c_4), \\ \qquad\qquad \text{if } \; 0 < x < t, 0 < y < t, t^2 < x^2 + y^2, \\ \frac{t}{2}(c_2 + c_4) + \frac{x}{2}(c_1 - c_2) + \frac{y}{2}(c_1 - c_4) + \frac{c_1 + c_3 - c_2 - c_4}{2\pi} J(t,x,y), \\ \qquad\qquad \text{if } \; 0 < x < t, 0 < y < t, x^2 + y^2 < t^2, \end{cases} \tag{5.32}$$

where

$$J(t,x,y) = \int_{\sqrt{x^2+y^2}}^{t} \frac{t}{\sqrt{t^2 - r^2}} \left(\frac{\pi}{2} - \sin^{-1}\frac{x}{r} - \sin^{-1}\frac{y}{r} \right) r dr.$$

It is obvious that $v \in L^\infty$. From the expression of v we know that v is a piecewise linear function outside the cone $x^2 + y^2 = t^2$. It is continuous on all characteristic surfaces and is C^∞ differentiable up to those characteristic surfaces. The only thing we have to give a careful analysis is the property of v in the inner side of the characteristic cone $x^2 + y^2 = t^2$. The property is determined by the function $J(t,x,y)$.

Denote $J_1(x,y) = J(1,x,y)$, it can be written as

$$\int_{x-\sqrt{1-y^2}}^{0} \int_{y-\sqrt{1-(x-\xi)^2}}^{0} \frac{d\xi d\eta}{\sqrt{1-(x-\xi)^2-(y-\eta)^2}}.$$

Then

$$\frac{\partial J_1}{\partial x} = \frac{\pi}{2} - \sin^{-1}\frac{y}{\sqrt{1-x^2}},$$

$$\frac{\partial J_1}{\partial y} = \frac{\pi}{2} - \sin^{-1}\frac{x}{\sqrt{1-y^2}},$$

$$\frac{\partial^2 J_1}{\partial x \partial y} = \frac{-1}{\sqrt{1-x^2-y^2}}.$$

In order to compute all derivatives of $J(t,x,y)$ we introduce the following lemma.

Lemma 5.7. *Let M denote any vector field in the basis (5.29), then for any multi-index α we have*

$$M^\alpha J = aJ + \sum_{|\beta|<|\alpha|} b_\beta M^\beta J_x + \sum_{|\beta|<|\alpha|} c_\beta M^\beta J_y, \tag{5.33}$$

where a,b_β,c_β are homogeneous functions of degree 0.

Temporarily leave the proof of Lemma 5.7 aside, we continue the proof of Lemma 5.6. Direct computation gives

$$\begin{aligned}
M_0 J_x &= M_1 J_x = 0,\\
M_2 J_x &= -(t^2 - x^2 - y^2)^{1/2}/(t+x),\\
M_0 J_y &= 0,\\
M_1 J_y &= t(t^2 - x^2 - y^2)^{1/2}/(t^2 - y^2),\\
M_2 J_y &= y(t^2 - x^2 - y^2)^{1/2}/(t^2 - y^2).
\end{aligned}$$

Since $t+x \neq 0$ and $t^2 - y^2 \neq 0$ in ω_1, and all $M_1 J_x, M_1 J_y$ are homogeneous functions of degree 0, then all of them are locally L^2 integrable. Moreover, in view of

$$\begin{aligned}
M_0(t^2 - x^2 - y^2) &= 2(t^2 - x^2 - y^2),\\
M_1(t^2 - x^2 - y^2) &= 0,\\
M_2(t^2 - x^2 - y^2) &= 0,
\end{aligned}$$

then continuously acting the operator M^α on $M_i J_x, M_i J_y$ always gives a homogeneous function of degree 0. This implies J_x, J_y are conormal distribution.

From the expression of J we also know $M^\alpha J \in L^2_{loc}(\omega_1)$ for any α. Similar discussion can be proceeded for other points on the characteristic cone. Hence $J(t,x,y) \in H^{0,k}$, so that $v \in H^{0,k}$ can be easily obtained. Lemma 5.6 is thus proved. □

Proof of Lemma 5.7. Let us prove Eq. (5.32) by induction. The equality is obvious for $|\alpha| = 1$. Now assume the equality is valid for $|\alpha| = k$, then for α_1 satisfying $|\alpha_1| = k+1$, we denote $M_1^\alpha = M^\alpha \tilde{M}$. Hence

$$\begin{aligned}
M_1^\alpha J &= M^\alpha(aJ + bJ_x + cJ_y)\\
&= \sum_{|\beta|\le|\alpha|} C_\alpha^\beta((M^\beta a)(M^{\alpha-\beta}J) + (M^\beta b)(M^{\alpha-\beta}J_x) + (M^\beta c)(M^{\alpha-\beta}J_y)).
\end{aligned}$$

According to the assumption of induction, substituting $M^{\alpha-\beta}J$ in the above equality by Eq. (5.32) gives us an expression of $M^{\alpha_1}J$ like a linear combination of right-hand side of Eq. (5.32) and the order of the operators acting on J_x or J_y is no more than k. Hence Eq. (5.32) is valid for all k. □

Next we are going to indicate $Ef \in B$. To this end we should compute the commutator of M^α and E.

Lemma 5.8. *If w is a function supported on $t \geq 0$, M satisfies $[\Box, M] = c_0 M$ with c_0 being a constant, then*

$$MEw = E(M + c_0)w. \tag{5.34}$$

Proof.

$$\begin{aligned}\Box MEw &= M\Box Ew + c_0\Box Ew \\ &= (M + c_0)w = \Box E(M + c_0)w.\end{aligned}$$

Then, by using the uniqueness of solution to Cauchy problem of the wave operator, we know MEw is equal to $E(M + c_0)$. □

We indicate here that the basis of tangential vector fields in ω_1 satisfies the commutate relations as

$$[M_0, \Box] = 2\Box,\ [M_1, \Box] = 0,\ [M_2, \Box] = 0. \tag{5.35}$$

Then Lemma 5.8 can be applied to estimate $Ef(u)$ in ω_1. However, in ω_5 we have

$$[N, \Box] = -2\partial_t(\partial_t + \partial_x + \partial_y),$$

which cannot be expressed as a linear combination of □ and the vector fields (5.30). Fortunately, there is not any characteristic direction in ω_5, then we can find some remidy to establish the necessary estimates for $Ef(u)$.

Lemma 5.9. *Assume that T_0 is sufficiently small, $u \in L^\infty \cap H^{0,k}(S, C_{T_0})$, and $\|u\|_{H^{0,k}(S,C_{T_0})} \leq G$, then $Ef(u)$ is also in $L^\infty \cap H^{0,k}(S, C_{T_0})$ and $\|Ef(u)\|_{H^{0,k}(S,C_{T_0})} \leq G$.*

Proof. The boundedness of $Ef(u)$ can be obtained from its explicit expression. In fact,

$$\begin{aligned}|Ef(u)| &\leq \frac{1}{2\pi}\int_0^t \int_{r \leq t-\tau} \frac{|f(u(\tau,\xi,\eta))|}{\sqrt{(t-\tau)^2 - (x-\xi)^2 - (y-\eta)^2}} d\tau d\xi d\eta \\ &\leq C\int_0^t \int_{r\leq t-\tau} \int_0^{2\pi} \frac{r}{\sqrt{(t-\tau)^2 - r^2}} d\theta d\tau dr \\ &\leq C' \int_0^t -\sqrt{(t-\tau)^2 - r^2}\Big|_0^{t-\tau} d\tau \leq C''.\end{aligned}$$

To prove the conormal regularity of $Ef(u)$, we use the open covering and related partition of unity as mentioned above. Write $Ef(u)$ as $\sum E(\chi_j f(u))$, we only have to prove $E(\chi_i f(u)) \in H^{0,k}$ for each i, and in fact we only have to consider the case $i = 1, 5$.

First let us consider $E(\chi_1 f(u))$, and estimate $\|M^\alpha E(\chi_1 f(u))\|_{L(\omega_j)}$ for each j. Since $M^\alpha = M_0^{\alpha_0} M_1^{\alpha_1} M_2^{\alpha_2}$ in ω_1, then by using Lemma 5.9 we have

$$\begin{aligned} M^\alpha(E\chi_1 f(u)) &= E((M_0+2)^{\alpha_0} M_1^{\alpha_1} M_2^{\alpha_2}(\chi_1 f(u)) \\ &= E \sum_{|\beta|\le|\alpha|} g_\beta M^{\beta_0}\chi_1 M^{\beta_1}u \cdots M^{\beta_\ell}u \end{aligned}$$

for $|\alpha| \le k$, where g_β and $M^{\beta_0}\chi_1$ are bounded. As indicated in Lemma 3.13, we have $M^{\beta_i u} \in L^{2k/|\beta_i|}$, and

$$\|M_i^\beta u\|_{L^{2k/|\beta_i|}(\omega_1\cap C_{T_0})} \le C\|u\|_{H^{0,k}(\omega_1\cap C_{T_0})},$$

then

$$\begin{aligned} &\|g_\beta M^{\beta_0}\chi_1 M^{\beta_1}u\cdots M^{\beta_\ell}u\|^2_{L^2(\omega_1\cap C_{T_0})} \\ &\le C\prod_{i=1}^{\ell} \|M_i^\beta u\|^2_{L^{2k/|\beta_i|}(\omega_1\cap C_{T_0})} \\ &\le C\|u\|^{2\ell}_{H^{0,k}(\omega_1\cap C_{T_0})} \le CG^{2\ell}. \end{aligned} \tag{5.36}$$

Combining it with the energy estimate satisfied by the operator E

$$\|Ef\|^2_{H^1(C_{T_0})} \le CT_0\|f\|^2_{L^2(C_{T_0})},$$

we obtain

$$\|M^\alpha(E\chi_1 f(u))\|_{L^2(C_{T_0})} \le \|M^\alpha(E\chi_1 f(u))\|_{H^1(C_{T_0})} \le C(G)T_0^{\frac{1}{2}}. \tag{5.37}$$

When $j \ne 1$, the basis of tangential vector field in ω_j can be expressed as linear combination of M_0, M_1, M_2. For instance, the basis in ω_2 is

$$M_0,\ M_1' = t\partial_y + y\partial_t,\ M_2' = x(\partial_t + \partial_y) + (t-y)\partial_x,$$

which can be written as

$$M_1' = \frac{t}{t-x}M_2 - \frac{y}{t-x}M_1, \quad M_2' = \frac{x}{t-x}M_2 + \frac{t-x-y}{t-x}M_1.$$

Similarly, in ω_3 the basis of tangential vector field is

$$M_0, \quad M_1'' = -t\partial_x - x\partial_t = -M_1,$$

$$M_2'' = y(\partial_t - \partial_x) + (t+x)\partial_y = \frac{t+x}{t-x}M_2 - \frac{2y}{t-x}M_1,$$

and in ω_5 the basis of tangential vector field is

$$M_0, \quad M_3 = M_1 - M_2,$$

$$N = \frac{t}{y^2 + x^2 - t^2}((y + x - t)M_0 + (y + x - t)M_1 + (y - x - t)M_2).$$

In view of $t - x \neq 0$ in ω_2, ω_3 and $t^2 - x^2 - y^2 \neq 0$ in ω_5, we can derive $E(\chi_1 f(u)) \in H^{0,k}$ by using the argument for the domain ω_1 as did above.

Next let us consider the conormal regularity of $E(\chi_5 f(u))$.

In ω_5 we can write $M_0^{\alpha_0} M_3^{\alpha_1} N^{\alpha_2} E(\chi_5 f(u))$ as

$$\sum_{\alpha_2+\beta_0+\beta_1 \leq |\alpha|} N^{\alpha_2} E(h_\beta M_0^{\beta_0} M_3^{\beta_1} u).$$

For $\alpha_2 = 1$, since E is a linear bounded operator from L^2 to H^1, then $E(h_\beta M_0^{\beta_0} M_3^{\beta_1} u) \in H^1$ and $N^{\alpha_2} E(h_\beta M_0^{\beta_0} M_3^{\beta_1} u) \in L^2$.

In the case $\alpha_2 > 1$, we use the equality

$$\begin{aligned} 2(t-x)(t-y)\Box &= (M_0^2 - M_0) - \left(M_3^2 - M_0 - \frac{t-x-y}{t}N\right) \\ &-2(x+y-t)\left(\frac{NM_0}{t} - \frac{N}{t}\right) + (t^2 - x^2 - y^2)\left(\frac{N}{t^2} - \frac{N^2}{t^2}\right). \end{aligned} \tag{5.38}$$

It implies

$$N^2 = a_0(t-x)(t-y)\Box + a_1 M_0^2 + a_2 M_3^2 + a_3 N M_0 + b_1 M_0 + b_2 N, \tag{5.39}$$

where a_i, b_i are homogeneous functions of degree 0 and bounded in ω_5. Hence, by denoting $\alpha_2 = \alpha_2' + 2$ and using Eq. (5.39), $N^{\alpha_2} E(h_\beta M_0^{\beta_0} M_3^{\beta_1} u)$ can be written as

$$N^{\alpha_2'} a_4 M_0^{\alpha_0} M_2^{\alpha_1} \chi_5 f(u) + \sum N^{\alpha_2''} E(h_\alpha M_0^{\alpha_0} M_2^{\alpha_1} u),$$

where $|\alpha_2'| = |\alpha_2| - 2, \alpha_2'' < \alpha, |a_4| = |a_0(t-x)(t-y)| \leq CT_0^2$. Continuing this process and applying Eq. (5.39) repeatedly the power of N can be reduced to 0 or 1. Hence by applying the conclusion on the case $\alpha_2 = 1$ we obtain

$$M_0^{\alpha_0} M_3^{\alpha_1} N^{\alpha_2} E(\chi_5 f(u)) \in L^2(\omega_5). \tag{5.40}$$

For $j \neq 5$, we can also construct the basis of tangential vector fields in ω_j by using of the linear combination of M_0, M_3, N, and then confirm $E(\chi_5 f(u)) \in H^{0,k}$ and corresponding energy estimate

$$\|E(\chi_5 f(u))\|_{H^{0,k}(C_{T_0})} \leq C(G) T_0^{1/2}. \tag{5.41}$$

We emphasize here once more that in the above proof $\chi_i = \chi_1$ and $\chi_i = \chi_5$ are two typical cases. As for other cases, the proof of the estimate $\|E(\chi_i f(u))\|_{H^{0,k}(C_{T_0})} \leq C(G) T_0^{1/2}$ is similar or even simpler. $\square$

Lemma 5.10. *The map $F[u] = v + Ef(u)$ is a contract map in $L^\infty \cap H^{0,k}$.*

Proof. By using the expression of E (Poisson integral), we know

$$\begin{aligned}\|F[u]-F[w]\|_{L^\infty} &= \|E(f(u)-f(w))\|_{L^\infty} \le CT_0\|f(u)-f(w)\|_{L^\infty} \\ &\le CT_0\|f'\|_{L^\infty}\|u-w\|_{L^\infty} \le \frac{1}{2}\|u-w\|_{L^\infty}.\end{aligned} \tag{5.42}$$

In order to estimate the $H^{0,k}$ norm of $F[u]-F[w]$, we use

$$\|F[u]-F[w]\|_{H^{0,k}} = \sum_i \sum_{|\alpha|\le k} \|M^\alpha(E\chi_i(f(u)-f(w)))\|^2_{L^2}. \tag{5.43}$$

Like what we did above, it is enough to consider the cases $i = 1, 5$. In the case $i = 1$ the basis of tangential vector fields can be chosen as (5.29), while in the case $i = 5$ the basis of tangential vector fields can be chosen as (5.30).

According to Eq. (5.32)

$$\begin{aligned}&\sum_{|\alpha|\le k} \|M^\alpha(E\chi_1(f(u)-f(w)))\|_{L^2(C_{T_0})} \\ &\le C\sum_{|\alpha|\le k} \|EM^\alpha(\chi_1(f(u)-f(w)))\|_{L^2(C_{T_0})} \\ &\le CT_0^{1/2}\sum_{|\alpha|\le k} \|M^\alpha(f(u)-f(w))\|_{L^2(C_{T_0})}\end{aligned}$$

Besides, direct computation gives

$$\begin{aligned}&\|M^\alpha(f(u)-f(w))\|_{L^2} \\ &\le \sum_{\ell\le k}(\|f^{(\ell)}(u)-f^{(\ell)}(w)\|_{L^\infty}\|M_1^\alpha u\cdots M^{\alpha_\ell}u\|_{L^2} \\ &\qquad + \|f^{(\ell)}(w)\|_{L^\infty}\|M^{\alpha_1}w\cdots M^{\alpha_\ell}(u-w)\|_{L^2}) \\ &\le C\|u-w\|_{L^\infty\cap H^{0,k}}.\end{aligned}$$

Hence

$$\|E(\chi_1(f(u)-f(w)))\|_{H^{0,k}} \le CT_0^{1/2}\|u-w\|_{L^\infty\cap H^{0,k}}. \tag{5.44}$$

The proof for the case $i = 5$ is similar to that case in the proof of Lemma 5.9. By using Eq. (5.39) we can decrease the power of N in the expression

$$M_0^{\alpha_0}M_3^{\alpha_1}N^{\alpha_2}E(\chi_5(f(u)-f(w))),$$

so that the power of N will be 0 or 1 finally. Notice that the operators M_0, M_3 can commutate with the operator E, then the energy estimate Eq. (5.36) implies

$$\|E(\chi_5(f(u)-f(w)))\|_{H^{0,k}} \le CT_0^{1/2}\|u-w\|_{L^\infty\cap H^{0,k}}. \tag{5.45}$$

Combining with Eq. (5.44) we obtain the contraction of the map E for sufficiently small T_0. □

Proof of Theorem 5.5. Define the sequence $\{u_k\}$ by

$$u_0 = v, \quad u_{k+1} = v + Ef(u_k). \tag{5.46}$$

Lemma 5.6 and Lemma 5.9 show that the sequence $\{u_k\}$ is well-defined. Lemma 5.10 implies the convergence of the sequence in $L^\infty \cap H^{0,k}$, and whose limit u is the solution of the problem (4.25) with the assigned conormal regularity. The contraction of the operator F gives the uniqueness of the problem (4.25).

Finally, we show that the singularity of u actually appear on all characteristic surfaces in (5.28). We will confirm it by showing that on these surfaces the possible singularity of $Ef(u)$ is weaker than the singularity of v.

On Σ_i $(1 \le i \le 4)$, the function v is continuous, and its derivatives of first order is discontinuous, hence the second derivatives of v have singularity like δ function supported on Σ_i, so that v is not in H^2_{loc}. For the surface Σ_5, v and its first derivatives are continuous on Σ_5, while the second derivatives of v goes to infinity like $d^{1/2}$ with d being the distance from the surface Σ_5, hence v is not in H^2_{loc} near Σ_5 either. On the other hand, since $f(u) \in L^\infty$, then $Ef(u) \in H^2_{loc}$. This means that the singularity of v is stronger than that of $Ef(u)$, so that v and $v + Ef(u)$ have singularity with same strength. This is the conclusion given in Theorem 5.5. □

Remark. The above method can also be applied to study the solution with flower-shaped singularities of general semilinear hyperbolic equations. However, in the case when the characteristic surfaces are not plane or cone, hence one may have to introduce some transformations to straighten surfaces and treat many lower order terms (see [40]). Besides, if the initial data of u itself has discontinuity then the solution u in $t > 0$ will also have stronger singularity.

5.3 Solutions with strong singularities of quasilinear equations (1-d case)

Comparing to the case for semilinear equation the study of the solutions with strong singularities for quasilinear equations is more complicated. The main reason is that the characteristics of equations depends on the unknown solutions. The fact has been emphasized before, when the solutions with weak singularities of quasilinear solutions are discussed in Chapter 4.

Let us start with the scalar quasilinear equation of first order.

$$\frac{\partial u}{\partial t} + u\frac{\partial u}{\partial x} = 0. \tag{5.47}$$

The problem on propagation of weak singularities of solutions to this equation can be discussed by using classical method or by using paradifferential operators. However, the propagation of strong singularities must be considered, because even a smooth solution of Eq. (5.47) may produce strong singularities. To see it we consider the Cauchy problem of Eq. (5.47) with the initial data

$$u(0, x) = \psi(x). \tag{5.48}$$

In a domain, where the solution u is C^1 regular, u takes constant on each characteristics $\frac{dx}{dt} = u$. It is easy to see, if $(0, \xi_1), (0, \xi_2)$ are two points on x-axis satisfying $\xi_1 < \xi_2$, and the function ψ satisfies $\psi(\xi_1) > \psi(\xi_2)$, then the characteristics issued from the points $(0, \xi_1)$ and $(0, \xi_2)$ must intersect somewhere. Indeed, these two characteristics are

$$x = \xi_1 + \psi(\xi_1)t, \quad x = \xi_2 + \psi(\xi_2)t, \tag{5.49}$$

which will intersect at the point P with coordinates $\left(\frac{\xi_2 - \xi_1}{\psi(\xi_1) - \psi(\xi_2)}, \frac{\psi(\xi_1)\xi_2 - \xi_1\psi(\xi_2)}{\psi(\xi_1) - \psi(\xi_2)}\right)$. Since a continuous function $u(t, x)$ cannot take two different values at a point P, then the solution u must have discontinuity before $t < \frac{\xi_2 - \xi_1}{\psi(\xi_1) - \psi(\xi_2)}$. Therefore, it is inevitable to study the solutions with discontinuity, if one wants to study Cauchy problem (5.47), (5.48) in a domain where t can be large.

For the solution u of Eq. (5.47), if u has jump on a curve Γ, which is transversal to x-axis, then $\frac{\partial u}{\partial x}$ is a δ function supported on Γ. In this case, the wave front set of u and $\frac{\partial u}{\partial x}$ can appear at a same point x with opposite direction. Then $u\frac{\partial u}{\partial x}$ in Eq. (5.11) is not well defined according to the regular definition of product of distributions. A reasonable treatment is to study the equation

$$u_t + \left(\frac{u^2}{2}\right)_x = 0 \tag{5.50}$$

instead of Eq. (5.47). Equation (5.50) is called equation with the form of conservation laws. For any C^1 solution, it is equivalent to the equation Eq. (5.47). Meanwhile, for any L^∞ function u, its square u^2 is well defined in classical way, then $\left(\frac{u^2}{2}\right)_x$ can be defined by the derivatives of distributions, so that the solution of Eq. (5.50) is well defined in the framework of distributions. Based on the form Eq. (5.50), if u satisfies

$$\iint \left(u\phi_t + \frac{u^2}{2}\phi_x\right) dxdt = 0 \tag{5.51}$$

for any C_0^∞ function ϕ, then u is called **weak solution** of Eq. (5.50). In accordance, if u satisfies

$$\iint_{t\geq 0} \left(u\phi_t + \frac{u^2}{2}\phi_x\right) dxdt + \int_{t=0} u_0\phi dx = 0 \tag{5.52}$$

for any finitely supported $C^\infty(t \geq 0)$ function ϕ, then u is called weak solution of Cauchy problem of Eq. (5.50) with initial data

$$u(0,x) = u_0(x). \tag{5.53}$$

Obviously, when $u \in C^1$, the weak solution u is the solution of Eq. (5.47) with Eq. (5.53) in classical sense.

For more general equation of conservation laws

$$u_t + (f(u))_x = 0, \tag{5.54}$$

the weak solution of its Cauchy problem with initial data

$$u(0,x) = u_0(x) \tag{5.55}$$

can be defined by

$$\iint_{t\geq 0} (u\phi_t + f(u)\phi_x)dxdt + \int_{t=0} u_0\phi dx = 0. \tag{5.56}$$

When u has jump on the curve $\Gamma : x = x(t)$, the conditions satisfying by the jump can be derived by using integral by parts. Indeed, let $[u], [f(u)]$ denote the jump of $u, f(u)$ on Γ, then for any C^∞ function ϕ

$$\int_\Gamma -\phi[u]dx + \phi[f(u)]dt = 0.$$

Hence at any point on Γ we have

$$s[u] = [f(u)], \tag{5.57}$$

where $[u] = u_r - u_\ell = u(t, x(t)+0) - u(t, x(t)-0)$, $[f(u)] = f(u_r) - f(u_\ell)$, s is the slope $\frac{dx}{dt}$ of Γ. The curve Γ with the slope s is called **shock**, while Eq. (5.57) is called **Rankine-Hugoniot condition**.

The slope of characteristics of Eq. (5.54) is $\frac{dx}{dt} = f'(u)$. When $f'' \neq 0$, one generally has $s \neq u_r, s \neq u_\ell$ if $u_r \neq u_\ell$. Therefore, the slope of shock is generally different from the slope of the characteristics on both sides of the shock. Such a phenomenon is different from the linear or semilinear case. It is also different from the propagation of weaker singularities.

An important fact is that for a given quasilinear equation one can give many different forms of conservation law corresponding to it. For instance, besides Eq. (5.50) we can also write

$$\left(\frac{u^2}{2}\right)t + \left(\frac{u^3}{3}\right)_x = 0, \tag{5.58}$$

which is equivalent to Eq. (5.50) for C^1 solutions. However, in the study of weak solutions the Rinkine-Hugoniot condition for Eq. (5.58) is

$$s\frac{1}{2}[u^2] + \frac{1}{3}[u^3] = 0, \tag{5.59}$$

which is different from Eq. (5.57). Therefore, in the study of weak solutions Eq. (5.50) is not equivalent to Eq. (5.58). In practice, we have to choose a suitable form of conservation laws. Fortunately, many partial differential equations are derived from physics, we may choose a right form, which describes the conservation laws in physics. Therefore, the form of conservation laws for a given differential system has been determined generally by its physical background.

In the study of discontinuous solutions for a given quasilinear equation one may also meet the difficulty on non-uniqueness. For instance, consider the Cauchy problem of Eq. (5.50) with the initial data

$$u(0,x) = \begin{cases} -1 & x < 0, \\ 1 & x > 0. \end{cases} \tag{5.60}$$

It is easy to check that the function

$$u_1(t,x) = \begin{cases} -1 & x < 0, \\ 1 & x > 0 \end{cases} \tag{5.61}$$

satisfies Eq. (5.50) and the initial condition (5.60) outside $x = 0$. Meanwhile, on $x = 0$ the solution satisfies the Rankine-Hugoniot condition. Besides, the function

$$u(t,x) = \begin{cases} -1 & x < -t, \\ x/t & -t < x < t, \\ 1 & x > t \end{cases} \tag{5.62}$$

also satisfies Eq. (5.50) in $t > 0$ and the initial condition (5.60) on $t = 0$. In fact, one can also write infinitely many functions with discontinuity, which satisfies Rankine-Hugoniot conditions on the lines bearing discontinuity, and satisfies the equation and initial condition outside these lines. Therefore, it is necessary to find a criterion to single out one weak solution to the Cauchy problem. Such a criterion is called **entropy condition**, under which the unique admissible weak solution is called **entropy solution**. For Eq. (5.50) the entropy condition is

$$\frac{u(t, x+a) - u(t, x)}{a} \leq \frac{E}{t}, \quad \forall\, a > 0, t > 0, \tag{5.63}$$

where E is independent of x, t, a. For Eq. (5.54) with $f''(u) > 0$, such a condition is also equivalent to that all characteristics issued from any point on shocks must be downward.

For quasilinear hyperbolic systems of conservation laws the entropy condition is also related to the distribution of characteristics. Next we restrict ourselves to the case of strictly hyperbolic systems.

Consider a hyperbolic system of conservation laws

$$\frac{\partial u}{\partial t} + \frac{\partial f(u)}{\partial x} = 0 \tag{5.64}$$

in domain $\Omega \subset R^2$, where $u, f(u)$ are vector functions with N components, $f(u)$ are C^∞ functions of u. The matrix $f'(u)$ has N real eigenvalues different from each other $\lambda_1(u) < \cdots < \lambda_N(u)$. Denote by $\lambda_i(u)$ the i-th eigenvalue, by $\mathbf{r}_i(u)$ the corresponding right eigenvector and by $\ell_i(u)$ the corresponding left eigenvector. If $\nabla\lambda_k \cdot \mathbf{r}_k \neq 0$, then the k-th eigenvalue is called **genuinely nonlinear**. Otherwise, if $\nabla\lambda_k \cdot \mathbf{r}_k \equiv 0$, then the k-th eigenvalue is called **linearly degenerate**.

Definition 5.1. If C^∞ function $w(u)$ satisfies

$$\langle \mathbf{r}_k(u), \nabla w(u) \rangle = 0, \tag{5.65}$$

then $w(u)$ is called k-th **Riemann invariant**. If in some domain the Riemann invariant is constant, then the corresponding solution u is called k-th simple wave. Moreover, if the simple wave is a function of $\dfrac{x - x_0}{t - t_0}$, then u is called **centered simple wave**.

Definition 5.2. If Γ is a smooth curve in Ω, the function u satisfies Eq. (5.64) on $\Omega \setminus \Gamma$. u has jump on Γ. At any point on Γ the slope of Γ and the value u_r, u_ℓ of u on both sides of Γ satisfy

$$f(u_r) - f(u_\ell) = s(u_r - u_\ell) \tag{5.66}$$

and

$$\lambda_{k-1}(u_\ell) < s < \lambda_k(u_\ell), \quad \lambda_k(u_r) < s < \lambda_{k+1}(u_r), \tag{5.67}$$

then Γ is called k-th **shock**. The conditions (5.66), (5.67) are called **Rankine-Hugoniot condition** and **entropy condition** respectively.

Definition 5.3. Assume that Γ is a smooth curve in Ω, the function u satisfies Eq. (5.64) on $\Omega \setminus \Gamma$ and has jump on Γ. Meanwhile, Γ is also characteristics corresponding to $\lambda_k(u_r)$ and $\lambda_k(u_\ell)$, then Γ is called k-th **contact discontinuity**.

We notice that the k-th contact discontinuity will appear only for the case when the k-th eigenvalue is linearly degenerate, and the k-th shock and k-th simple wave as defined in Definitions 5.1 and 5.2 will appear only for the case when the k-th eigenvalue is genuinely nonlinear.

Consider the Cauchy problem of Eq. (5.64). If the initial data has jump, then the solution may also has jump. Particularly, when the initial data are taken as

$$u(0,x) = \begin{cases} u_\ell & x < 0 \\ u_r & x > 0, \end{cases} \tag{5.68}$$

then the corresponding Cauchy problem is called **Riemann problem**. For such a Riemann problem we have the following result (see [84]).

Theorem 5.6. *Assume that the system Eq. (5.64) is strictly hyperbolic, its characteristics are genuinely nonlinear or linearly degenerate, $|u_r - u_\ell|$ is sufficiently small, then in a neighborhood of the origin there is a unique entropy solution, which depends on x/t, and contains at most N nonlinear waves (shocks, centered waves and contact discontinuity). The solution is constant between these nonlinear waves.*

We notice that the contact discontinuity itself is characteristics, where the solution has discontinuity. The front and the back of the center wave are also characteristics, where solution is continuous, and its derivatives have discontinuity. Therefore, from the viewpoint of singularity analysis the above theorem indicates that the strong singularity of the initial data will propagate along shocks and characteristics into the domain $t > 0$. The path of propagation of singularities forms a fan-shaped figure.

When the initial data are taken as more general functions with jump like

$$u(0,x) = \begin{cases} u_-(x) & x_- < x < 0 \\ u_+(x) & x_+ > x > 0, \end{cases} \tag{5.69}$$

we may consider the system Eq. (5.54) in the domain

$$T(\delta) = \{(t, x); x_- + At \le x \le x_+ - At\},$$

where $A > \sup(|\lambda_k(u)|)$. For such Cauchy problems with discontinuous initial data the following local existence theorem holds.

Theorem 5.7. *Assume that the system Eq. (5.64) is strictly hyperbolic, its characteristics are genuinely nonlinear or linearly degenerate, the functions $u_-(x), u_+(x) \in C^{2,1}$, and $|u_-(0) - u_+(0)|$ is sufficiently small, then there is $\delta_0 > 0$, such that the Cauchy problem admits an L^∞ unique entropy solution in $T(\delta_0)$. The solution contains at most N nonlinear waves (shocks, centered waves and contact discontinuity). The solution is $C^{2,1}$ smooth between these nonlinear waves.*

For the proof of the theorem we refer readers to [88].

Based on the above local results it is possible to study the global property of solutions including the propagation of nonlinear waves. However, since in most cases the results on description of singularities are related to the existence of solutions with given singularities, there is not a general result on propagation of strong singularities for quasilinear equation like that on propagation of weak singularities. Here we will not list specific results on it, and prefer to turn to the discussion on multidimensional case.

5.4 Solutions with strong singulari ties of quasilinear equations (m-d case)

5.4.1 *Fan-shaped singularity structure*

The system of conservation laws in multidimensional space can be written as

$$\frac{\partial u}{\partial t} + \sum_{i=1}^{n} \frac{\partial F_i(u)}{\partial x_i} = 0, \tag{5.70}$$

where n is the dimension of the space dimension. As we did in Section 3 we will only introduce some results on the existence and the propagation of singularities of solutions and refer readers to related references for their proof.

In [92] A.Majda started the study of weak solutions to multidimensional system of conservation laws with fan-shaped wave structure. He considered the Cauchy problems of Eq. (5.70) with initial data

$$u(0, x) = u_0(x), \tag{5.71}$$

where $u_0(x)$ has jump on a curve Γ on the initial plane $t=0$, and proved the following theorem.

Theorem 5.8. *Assume that the coefficients of (5.70), and the initial data (5.71) satisfy:*

(1) The system (5.70) is strictly hyperbolic, $F_i(u)$ are C^∞ functions of their argument, and there is a positive symmetric matrix $A_0(u)$, such that all $A_0(u)F_i'(u)$ are symmetric.

(2) $u_{0-}(x)$,$u_{0+}(x)$ belong to H^{s+1} on the both sides of Γ with integer $s>2[\frac{n}{2}]+7$.

(3) Denote by α the parameter of the curve Γ, whose equation $\sigma(\alpha)\in H^{s+1}$ and satisfies

$$-\sigma(\alpha)(u_{0+}(\alpha)-u_{0-}(\alpha))+\sum n_i(\alpha)(F_i(u_{0+}(\alpha))-F_i(u_{0-}(\alpha)))=0, \quad (5.72)$$

and the compatibility conditions up to $(s-1)$ order on Γ. Moreover, there is an integer p, such that

$$\lambda_p(u_{0-}(\alpha))>\sigma(\alpha)>\lambda_p(u_{0+}(\alpha)),$$

$$\lambda_{p+1}(u_{0+}(\alpha))>\sigma(\alpha)>\lambda_{p-1}(u_{0-}(\alpha)).$$

(4) For any point on Γ, the initial problem with constant coefficients defined by $(u_{0+}(\alpha),u_{0-}(\alpha),\sigma(\alpha))$ is uniformly stable.

Then for sufficiently small T, in the domain $0<t<T$ there is an H^{s+1} surface S and H^s functions $u_-(t,x),u_+(t,x)$, satisfying (5.70), (5.71) and Rankine-Hugoniot conditions and entropy condition on S.

In the statement of the above theorem the compatibility conditions means that the derivatives of $u_\pm$ on the both sides of Γ satisfy the equalities derived from differentiation of Rankine-Hugoniot conditions. Besides, in [92] the Cauchy problem with discontinuous data is treated as a free boundary value problem, and the shock front is the free boundary, which is to be determined together with the solution. In the free boundary value problem the Rankine-Hugoniot conditions give a differential relation of σ and the solution $u_\pm$ on both sides. An important fact is that the relation is an elliptic differential system for σ. Meanwhile, as a boundary value problem for the original system of conservation laws the uniform stability condition amounts to the usual Lopatinski condition for initial boundary value problems of hyperbolic systems. These crucial points enable us to establish suitable estimates dominating the variation of the solution with a shock front. Then the estimates directly lead us to the stability of the solution to the linearized problem and the existence of the local solution of the nonlinear problem near the curve Γ.

Theorem 5.8 indicates that under suitable assumptions the singularity in the initial data will only propagate along a shock front issued from Γ. Furthermore, for the propagation of singularity (or regularity) along shock front we proved in [34]

Theorem 5.9. *Assume that u is a solution with shock front S of the system (5.70) in Ω, which is located in the determinacy domain of $\Omega_- = \Omega \cap \{t < 0\}$. Let $s > 1/2$, assume that in Ω the shock front S is an H^{s+1} surface, and $u_\pm \in H^s$ in the both sides of S. Then $S \in C^\infty$ and $u_\pm \in C^\infty$ in Ω_- imply $S \in C^\infty$ and $u_\pm \in C^\infty$ in the whole Ω.*

Under some other restrictions on the initial data one can find a solution to the Cauchy problem containing a single multidimensional centered rarefaction wave. For the convenience of presentation we assume that the space-dimension is 2, and denote (x_1, x_2) by (x, y). Then Eqs. (5.70) and (5.71) become

$$\begin{cases} \dfrac{\partial u}{\partial t} + \dfrac{\partial f(u)}{\partial x} + \dfrac{\partial g(u)}{\partial y} = 0. \\ u(0, x, y) = u_0(x, y). \end{cases} \tag{5.73}$$

For the Cauchy problem (5.73), we assume that the system is strictly hyperbolic, $\lambda(u, \eta)$ is the eigenvalue of $f'(u) + \eta g'(u)$, and the system is genuinely nonlinear with respect to the eigenvalue. Assume that the initial data of u has discontinuity on the curve Γ, then by using the value of u on the left side and the right side of Γ we can determine two characteristic surfaces $\pi_\pm : x = \psi_\pm(t, y)$ issuing from Γ. The centered rarefaction wave is such a solution $u(t, x, y)$, which is continuous on $t \geq 0$ except Γ, differentiable on $t > 0$ except $\pi_\pm$, satisfying the system (5.73) in $t > 0$ and satisfying the initial data on $t = 0$. Since $\pi_\pm$ is characteristic, then on $\pi_\pm$ the following relation holds:

$$(\psi_\pm)_t(t, y) = \lambda(u_\pm(t, \psi_\pm(t, y), y), -(\psi_\pm)_y(t, y)). \tag{5.74}$$

Moreover, there are a set of characteristic surfaces $s(t, x, y) = const.$ passing through Γ, such that $u(t, x, y)$ takes different limit as (t, x, y) tends to Γ along different characteristic surface. All characteristic surfaces issuing from Γ likes a fan with a front surface π_+ and a back surface π_-. The front surface and the back surface are unknown and has to be determined with the solution together, hence they can also be regarded as free boundaries.

Different from the shock front solution case the front and the back of the centered rarefaction wave are characteristics.

According to the character of u, we can write u as $H(s(t,x,y),t,x,y)$, where $s(t,x,y)$ satisfies

$$(s_t + A_1 s_x + A_2 s_y)H_s' = 0, \tag{5.75}$$

and ∇s is unbounded near Γ. Equation (5.75) means that H_s' is the right eigenvector r of the matrix $A_1 s_x + A_2 s_y$. Therefore, the solution $h(s,y)$ of the initial value problem

$$\frac{dh}{ds} = r(h, -\phi_0'(y)),\ h(0,y) = u_0^+(\phi_0(y),y), \tag{5.76}$$

satisfies $H(s,0,0,y) = h(0,y)$.

From the above description of the centered rarefaction wave we see that in order to find a centered rarefaction wave to connect the solution $u_-(t,x,y)$ and $u_+(t,x,y)$, the initial data should admit a function $s(y)$, such that

$$h(s(y),y) = u_{0-}(\phi_0(y),y). \tag{5.77}$$

It is also called compatibility condition. Furthermore, If one can compute the values of all k-th derivatives of $u(t,x,y)$ in piecewise smooth domains near Γ by applying (5.73) and the compatibility condition (5.77), such that these values can determine an approximate solution satisfying (5.73) with error $O(t^{k+1})$, then we say the compatibility condition of k-th order is satisfied.

Theorem 5.10. *For the Cauchy problem (5.73) and a given real number $s > 0$, assume that the coefficients and the initial data satisfy the following conditions.*

(1) $f(u), g(u)$ are C^∞ functions of their argument. Equation (5.71) is strictly hyperbolic system, and there is a positive symmetric matrix $A_0(u)$, such that $A_0(u)f'(u), A_0(u)g'(u)$ are symmetric.

(2) $u_{0-}(x,y)$ ($u_{0+}(x,y)$ resp.) belongs to H^k in the domain left (right resp.) to Γ, where k is a suitably integer larger than s.

(3) k-th compatibility condition on Γ satisfied, where d is a positive number determined by the coefficients of Eq. (5.73).

Then there is a solution of the Cauchy problem with a centered rarefaction wave in a neighborhood of Γ. The solution belongs to H^s outside Γ and the front and back characteristic surfaces of the centered wave.

Slightly change the statement of the theorem we can obtain the conclusion for the corresponding problem in the space $(t, x_1, \cdots, x_n)$.

The proof of Theorem 5.10 can be found in [5]. In the proof the author applied a suitable weighted Sobolev space to measure the regularity of approximate solutions and the Nash-Moser iterative scheme to avoid the "derivative loss" in the process of iteration. This theorem means that under suitable conditions the singularity of the initial data will propagate with weaker strength along two characteristic surfaces issuing from Γ.

For the Cauchy problem (5.73), the initial data may also only have weak singularity on a curve Γ, for instance, the initial data are continuous but their derivatives of some order are discontinuous. In this case, the singularity will propagate along characteristic surfaces issuing from Γ. Such waves carrying weak singularities are called **sound waves**. The existence and propagation of sound waves were analyzed in [105].

The more difficult case in studying Cauchy problem of multi-dimensional system of conservation laws with discontinuous initial data is the case of contact discontinuity, which is a characteristic surface formed by stream lines. Different from the centered rarefaction wave case the solution on this characteristic surface is also discontinuous. On the other hand, different from the shock front case the uniform Kreiss-Lopatinskii condition on the boundary is not satisfied. In other words, in the study of multidimensional contact discontinuities one has to deal with the difficulties appearing in both the shock front case and the rarefaction wave case. J-F.Coulombel and P.Secchi studied the vortex sheets case for multidimensional system of conservation laws in [58], where the contact discontinuity is also called compressible vortex sheet. For the Cauchy problem of Euler system they proved the existence of solution with contact discontinuity under an additional "supersonic condition". The details can be found in [58].

When the initial data do not satisfy the restrictions given in the above three special cases the solution to the multidimensional systems under consideration may contain more nonlinear waves, like two shocks [102], [107], one shock and one rarefaction wave [86], one shock and one sound wave [42], two sound waves [52] etc. General data with discontinuity on a smooth curve may develop all three kinds of nonlinear waves (shock, rarefaction wave and contact discontinuity). Although the main difficulties in the three individual cases have been overcome, so far the result in most general cases has not been established yet. Obviously, such a result is significant and is anticipated.

Fan-shaped wave structures appear in many other problems of multidimensional hyperbolic systems. In [36] the author studied the reflection of a moving shock hitting a smooth wall. If the moving shock is a planar shock and the wall is a convex surface, then near the touch point the local existence of the solution describing the reflection of the shock is proved. However, the reflection phenomena will be greatly complicated when time goes on: various irregular reflection of shocks may happen (e.g. see [17], [48]).

The interaction of shocks was discussed in [103]. For 2×2 quasilinear hyperbolic system the interaction of shock waves will only produce two new shocks. The situation can be explained as each shock runs across another one and propagates further. But for the system with more than two equations (and unknown functions) the interaction of two shocks may produce more singularities. Generally, the problem can be reduced to a Cauchy problem with discontinuous data, and then treated as a corresponding Cauchy problem.

An interesting and important problem in gas dynamics is the study of supersonic flow past a wedge. For a steady supersonic flow past a three-dimensional wedge, when the attack angle and the vertex angle of the wedge are well controlled (for instance, the sum of the attack angle and the vertex angle is less than a critical value), an attached shock front at the edge of the wedge will be formed. The physical problem to determine the location of the shock front and the flow behind the shock can be reduced to a boundary value problem in a domain between the attached shock front and the surface of the wedge. The shock front is the free boundary for this boundary value problem. The local existence of the solution with the attached shock front was proved in [39]. The global existence of such solutions and the propagation of shock fronts were given in [148].

5.4.2 *Flower-shaped singularity structure*

It should be emphasized that in the multidimensional space many complicated wave structure are not 1-D like structure with M-D perturbations. Such structures are essentially multidimensional. Based on the progress in the study of various multidimensional problems with fan-shaped wave structure people are enable to pay more attention to the study of the essentially multidimensional problems.

A good example of the essentially multidimensional problem for quasilinear hyperbolic systems is the problem on supersonic flow past a pointed body. Like the problem on supersonic flow past a wedge, when the vertex angle of the pointed body is less than a critical value, the shock front is attached at the tip of the body, forming a bigger conical surface. Obviously, such a shock front is not a perturbation of a plane shock, and the state between the shock and the surface of the body is not a perturbation of a constant state either. Here the shock front and the surface of the body are two conical surfaces issuing from a single point – the tip of the conical body. Hence such a wave structure is **flower-shaped wave structure**. We notice that the boundary of the domain has strong singularity at the tip, which causes new difficulty in seeking the solution.

In [45] the author gives a proof of the existence of shock front solution near the tip. The problem is first approximated by the straight version of the original problem, i.e. the pointed body is replaced by a conical body with straight generating lines, and the coming flow is assumed to be constant. Then the second approximation is applied to treat the perturbation of the straight version of the problem.

Let (R, θ, z) be the regular cylindrical coordinates, $r = R/z$, and the surface of a conical body is described by $r = b(z, \theta)$. The main result obtained in [45] is

Theorem 5.11. *Assume that $b(z, \theta)$ is a small perturbation of a constant b_0, which satisfying*

$$\|b(0,\theta) - b_0\|_{C^{k_1}} \leq \epsilon_0,$$

$$\partial_Z^k b(0,\theta) = 0 \quad \text{for} \quad 1 \leq k \leq k_2,$$

where k_1, k_2 are integer determined by the conical surface and the data of the problem. Assume that a supersonic flow parallel to the z-axis comes from infinity with speed $q = q_\infty$ satisfying $q_\infty > a_\infty$ ($= (\frac{\gamma p_\infty}{\rho_\infty})^{\frac{1}{2}}$), where p_∞, ρ_∞ are the pressure and the density at infinity respectively. Besides, b_0 is less than a critical value b_* determined by $q_\infty, p_\infty, \rho_\infty$. Then the problem of the supersonic flow past the pointed body admits a local weak entropy solution with a pointed shock front attached at the origin.

The result on the existence of the solution with its shock front structure near the tip of the body enable us to study global existence and the asymptotic behavior of the flow behind the shock waves (see [55], [90], [139] etc.).

From the viewpoint of singularity propagation the phenomenon described in the above theorem can be understood in the following way: the singularity of the boundary of the domain produces strong singularity of the solution, and the assigned singularity structure describes the manner of propagation of singularities.

In the study of unsteady compressible flow the flower-shaped wave structure appears more frequently. For instance, consider two-dimensional Riemann problem, which is a Cauchy problem of the two-dimensional system of conservation laws

$$\frac{\partial u}{\partial t} + \frac{\partial f(u)}{\partial x} + \frac{\partial g(u)}{\partial y} = 0 \tag{5.78}$$

with piecewise constant initial data, which takes different constants in different angular domains. Since both the system (5.78) and the initial data are invariant under the dilation of coordinates

$$t \to \alpha t, \quad x \to \alpha x, \quad y \to \alpha y,$$

then we may consider the solution of 5.78) with the form

$$u(t,x,y) = v(x/t, y/t),$$

which is called self-similar solution. Obviously, if $v(\xi, \eta)$ has some singularity on a curve $\gamma: \ g(\xi, \eta) = 0$, then in the space (t, x, y) we can find a curve generating from the origin and gradually grows up keeping its shape resemble to γ. Hence v has a flower-shaped wave structure. In general, a d+1 dimensional unsteady problem (1 time-dimension plus d space-dimensions) may produce a d-dimensional problem in self-similar coordinate system independent of t. Hence the latter is usually called **pseudo-steady problem**.

Far away from the origin the influence of the origin vanishes, so that the Cauchy problem is one space-dimensional, which can be solved by using the theory of one-dimensional system of conservation laws. In accordance, for a given two-dimensional Riemann problem one may have many nonlinear waves (formed by straight characteristics) coming from infinity in different directions, and these nonlinear waves will interact when they meet together. The plentiful phenomena of interaction of these waves lead to the great complexity of the nonlinear wave structure either in the self-similar coordinates or in the original physical coordinates.

Like the setting of the Riemann problems we can also consider some initial boundary value problems, which are invariant under dilation of time coordinate and space coordinates. Such problems are called **initial boundary value problems of Riemann type**. Many physical problems can be derived in such a way, that the initial data take different constants in different sectors, while some sectors are solid, where no flow could go into.

For instance, we can take initial data as follows. The whole plane is separated by the rays $\theta = \theta_0$ $(0 < \theta_0 < \pi/2)$, $\theta = \pi/2$ and $\theta = -\pi$ to three sectors. The sector $-\pi < \theta < \theta_0$ are solid and no gas can go into. Meanwhile, the gas is assumed to take different constant states in $\theta_0 < \theta < \pi/2$ and $\pi/2 < \theta < \pi$. Moreover, the flow parameters in both sides of $\theta = \pi/2$ can determine a single plane shock moving forward to the ramp $\theta = \theta_0$. Then, the initial boundary value problem with such data can be regarded as a moving planar shock with constant speed hits a ramp at $t = 0$. Such a problem is called "shock reflection by a ramp".

The problem "shock reflection by a ramp" is a typical problem in nonlinear hyperbolic system with great important effect in physics. It offers many possibility of complicated wave structure including regular reflection, Mach reflection and other kind irregular reflection. All these structures are self-similar in (t, x, y) space, so that are flower shaped. For this problem we refer readers to [17], [47], [48], [57] and references therein.

Like the above setting one can also discuss other initial boundary value problems of Riemann type. For instance, in the above example, if the flow parameters on the both side of $\theta = \pi/2$ can determine a single rarefaction wave moving forward to the ramp, then we obtain a problem "reflection of rarefaction wave by a ramp". Similarly, if $\theta_0 < 0$ we can obtain the problem "shock diffraction by a convex angle". In the latter case there may not be any reflected shock, but there is a sonic wave propagating from the origin to infinity.

Another interesting Riemann type initial boundary value problem is "damp collapse". Assume that there is a reservoir having wedge shape filled with static water. By using the polar system to describe the plane R^2, we may assume that the damp of the reservoir is $\theta = \pm\theta_0$, and the water is stored in the domain $-\theta_0 \leq \theta \leq \theta_0$. If at the time $t = 0$ the damp of the reservoir suddenly collapses, then the water floods out of the reservoir. By taking the hight of the water as the unknown function, which satisfies the 2-d shallow water equation. The front of the water is the singularity of the height function. Since the shallow water equation has much similarity with the Euler equation in gas dynamics, the above problem can also be understood as the problem of expansion of gas into vacuum. Such a problem is first put forward by [129]. Recently, [91] gives a proof of the existence of solution to the problem and described the singularity structure of the solution.

Chapter 6

Formation of shocks for quasilinear hyperbolic equations

As indicated in the previous chapters, singularities of solutions for linear partial differential equations propagate along the whole characteristics, and the weak singularities of solutions for nonlinear equation also propagate along characteristics like this. However, the evolution of a smooth solution of a quasilinear equation may produce new strong singularities. How does strong singularity grows out from a smooth solution? The process of the formation of singularities is obviously an interesting and important problem, which is the main object in this Chapter.

6.1 The case of scalar equation

6.1.1 *Two mechanism of blow-up of smooth solutions*

It is a characteristic phenomenon for nonlinear equation, that a smooth solution may develop singularities. The phenomenon of forming singularities from a smooth solution is called **blow-up** of the solution. There are basically two mechanism of the formation of singularities. One is that the value of the solution tends to infinity, the other is that the derivative of the solution tends to infinity though the value of the solution is bounded. To explain such phenomena we introduce two examples for scalar equations.

Consider the Cauchy problem of a semilinear equation of first order

$$\begin{cases} \dfrac{\partial u}{\partial t} + \dfrac{\partial u}{\partial x} = u^2, \\ u(0,x) = u_0(x), \end{cases} \tag{6.1}$$

where $u_0(x)$ is a smooth function. It is easy to know that for small t there is a smooth solution of the problem (6.1). The solution can be expressed explicitly as

$$u(t,x)=\frac{u_0(x-t)}{1-tu_0(x-t)}. \tag{6.2}$$

Obviously, if the initial data u_0 is positive at a point ξ, then along the characteristics $x-t=\xi$, the solution $u(t,x)$ increases as t increases, and the solution does not make sense as $t=1/u_0(\xi)$. Therefore, on the (t,x) plane there is a curve defined by $tu_0(x-t)=1$, above which the solution does not exist. Taking $M=\max u_0(\xi)$. If $M>0$, the global smooth solution of the problem (6.1) exists only for $t<M^{-1}$. Such a mechanism of blow-up of smooth solution is similar to the blow-up of solution to the ordinary differential equation $\frac{dy}{dx}=y^2$, so that is called **ode mechanism**.

Another mechanism can be figure out from the Cauchy problem of a quasilinear equation. Consider

$$\begin{cases}\dfrac{\partial u}{\partial t}+\dfrac{\partial f(u)}{\partial x}=0,\\ u(0,x)=u_0(x),\end{cases} \tag{6.3}$$

where $f(u)$,$u_0(x)$ are smooth functions. Then near $t=0$ the problem (6.3) can be solved by using the characteristic method. Since the solution u takes constant in each characteristics, then $u(t,x)$ must be bounded, as $u_0(x)$ is bounded. However, if $f''(u)>0$ and $u_0(x)$ is decreasing, then the derivative of u will go to infinity as t increases.

Indeed, by taking derivatives with respect to x for (6.3), we have

$$\frac{\partial u_x}{\partial t}+f'(u)\frac{\partial u_x}{\partial x}=-f''(u)u_x^2.$$

Denote by d/dt the differentiating along the characteristics $x=x(t)$, we have

$$\frac{du_x}{dt}=-f''(u)u_x^2.$$

By integrating we obtain

$$\frac{1}{u_x}-\left(\frac{1}{u_x}\right)_0=\int_0^t f''(u)d\tau,$$

then

$$u_x(t,x(t))=\frac{u_0'(x(0))}{1+u_0'(x(0))\int_0^t f''(u)d\tau}. \tag{6.4}$$

Obviously, if $f''(u)>0, u_0'(x(0))<0$, then the denominator of (6.4) decreases, as t increases. The value $|u_x|$ can even increase infinitely along the characteristics. Hence under the assumption $f''(u) > 0$, if $u_0'(\xi) < 0$, then there is a $t(\xi)$, such that along the curve γ the derivative of u tends to infinity as $t \to t(\xi)$. Denote $t^* = \min\{t(\xi)\}$, we find that the global smooth solution of the Cauchy problem (6.3) exists as $t < t^*$. Corresponding to t^*, there is a characteristics $\gamma : x = x(t)$, such that the derivative of u tends to infinity as $t \to t^*$. Denote $x^* = x(t^*)$, the point (t^*, x^*) gives the time and location of blow-up of the solution u. The mechanism of blow-up, which let all characteristics squeeze together to produce the infinity of derivatives, is called **geometric mechanism**. Since the characteristics of semilinar equations does not depend on solutions, then geometric mechanism does not appear for semilinear equations. While for quasilinear equations or fully nonlinear equations both mechanism of blow-up are possible.

It is worth indicating that for Cauchy problem (6.3) the case, when the solution itself and its derivatives of first order are bounded but the derivatives of second order blow up, is impossible. Indeed, by differentiating Eq. (6.3) with respect to x twice we obtain

$$\frac{\partial u_{xx}}{\partial t} + f'(u)\frac{\partial u_{xx}}{\partial x} + 3f''(u)u_x u_{xx} = -f^{(3)}(u)u_x^3. \tag{6.5}$$

Obviously, this is a linear equation of u_{xx}. If u and u_x are continuous and bounded, then all coefficients and the right hand side are bounded. Therefore, the initial boundedness of u_{xx} ensure its boundedness for all time.

Although the above discussion is doing for scalar nonlinear differential equations of first order, people believe that for differential equations of higher order or for systems there are also these two mechanism of blow-up mainly. If the equation is m-th order, then the ode mechanism corresponds to the case when the solution or its derivatives of order lower than m tend to infinity. Conversely, the geometry mechanism corresponds to the case when all derivatives of order lower than m is bounded and the derivatives of order m goes to infinity. In this chapter we are more interested in discussing geometric mechanism, particularly, the method to construct a solution with singularity in a neighborhood of the blow-up point.

6.1.2 *Formation of a shock*

The above discussion indicates that the smooth solution of Eq. (6.3) will blow up at the point $P(t^*, x^*)$. More precisely, we can prove that the a shock is produced from P, whose strength increases gradually starting from zero. Next we are going to construct a solution $u(t, x)$, which is a

C^1 solution for $t < t^*$, but with discontinuity on a curve L starting from P. The value of $u(t,x)$ on both sides of L satisfies the Rankine-Hugoniot condition. To this end we first give the construction of $u(t,x)$, then give the precise theorem and its proof.

Denote $g(y) = f'(u_0(y))$, the family of the characteristic lines can be expressed by

$$x = y + tg(y). \tag{6.6}$$

Denote by $y(t,x)$ the point of the intersection of the characteristics passing (t,x) with x-axis, then

$$x = y(t,x) + tg(y(t,x)). \tag{6.7}$$

For small t the function $y(t,x)$ can be uniquely determined by the implicit theorem, hence the solution of Eq. (6.3) is

$$u(t,x) = u_0(y(t,x)) \tag{6.8}$$

for small t. When $g'(y) < 0$, the family of characteristics forms an envelop Γ as t increases. The envelop Γ is determined by Eq. (6.6) and

$$1 + tg'(y) = 0. \tag{6.9}$$

Equation (6.9) gives $t = -1/g'(y)$, then the coordinates of the tangential point of the envelop with the corresponding characteristics is $(-1/g'(y), y - g(y)/g'(y))$. The maximum of $-1/g'(y)$ corresponds to the cusp of the envelop. Without loss of generality we only consider the case there is only one cusp on the envelop. Hence we assume that there is a y_0, such that

$$(y - y_0)g''(y) > 0. \tag{6.10}$$

Obviously, under the assumption (6.10), $1/g'(y)$ takes minimum t_0 at y_0. In accordance, $(t_0, y_0 + t_0 g(y_0))$ is the cusp of the envelop.

When $t > t_0$, Eq. (6.9) has two roots $y = \eta_-(t)$ and $y = \eta_+(t)$ $(\eta_-(t) < y_0 < \eta_+(t))$. Correspondingly, two branches of Γ are $(t, x_\mp(t))$ with $x_\mp(t) = \eta_\mp(t) + tg(\eta_\mp(t))$. Denote by G the domain between these two branches, then $G = \{x_+(t) < x < x_-(t), t > t_0\}$. Outside G the characteristics corresponding to point (t,x) is unique, then the solution u can be expressed by Eq. (6.8). However, there are three roots $y_-(t,x) < y_c(t,x) < y_+(t,x)$ for each point (t,x) in G. This means that when the foot of a characteristic line moves from $-\infty$ to ∞, each point (t,x) in G is swept by the characteristics three times. Therefore, the solution in G cannot be expressed only by Eq. (6.8).

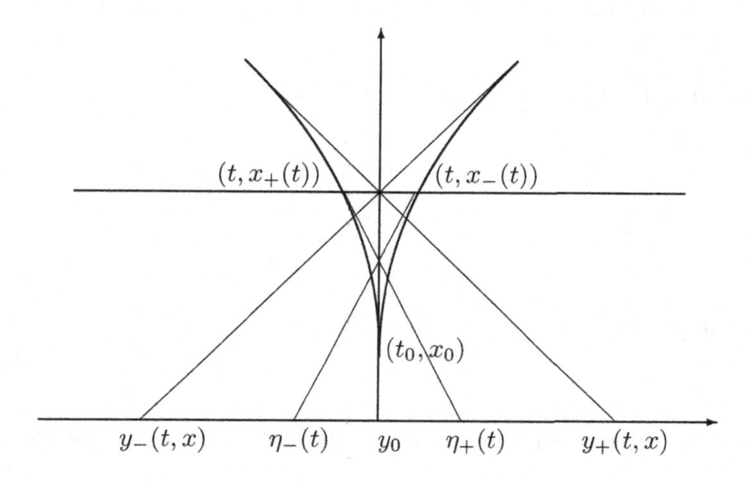

Figure 6.1. Characteristics and their envelope near the cusp

Denote

$$\Omega_+ = \{t \geq t_0, x > x_+(t)\},$$

$$\Omega_- = \{t \geq t_0, x < x_-(t)\},$$

we can define

$$u_\pm(t,x) = u_0(y_\pm(t,x))$$

in $\Omega_\pm$. Obviously, both $u_\pm(t,x)$ are well defined in $G = \Omega_- \cap \Omega_+$. Making an initial value problem

$$\begin{cases} \phi'(t) = \dfrac{f(u_+(t,\phi(t))) - f(u_-(t,\phi(t)))}{u_+(t,\phi(t)) - u_-(t,\phi(t))}, \\ \phi(t_0) = x_0. \end{cases} \tag{6.11}$$

If we can prove that there is a unique solution $\phi(t)$ of Eq. (6.11) in G, and $\phi(t)$ satisfies $x_-(t) < \phi(t) < x_+(t)$, then we can define

$$u(t,x) = \begin{cases} u_+(t,x), & x > \phi(t), \\ u_-(t,x), & x < \phi(t), \end{cases} \tag{6.12}$$

which is the weak solution of (6.3). Indeed, Eq. (6.11) is nothing but the Rankine-Hugoniot condition of any shock. Moreover, since the direction of characteristics on both sides of Γ is downward, then the entropy condition is also satisfied. Therefore, the theorem of existence of solution to (6.3) can be expressed as follows.

Theorem 6.1. *Assume that* $f(u) \in C^\infty, u_0(y) \in C^p (p \geq 4)$, *and* $g(y) = f(u_0(y))$ *satisfies*

$$g'(y_0) < 0,\ (y - y_0)g''(y) > 0,\ g^{(3)}(y_0) > 0,$$

then by denoting

$$t_0 = -\frac{1}{g'(y_0)},\ x_0 = y_0 + t_0 g(y_0),$$

the problem (6.11) has C^1 *solution in* $t \geq t_0$, *which belongs to* C^p *in* $(t_0, +\infty)$. *Correspondingly, (6.12) is the weak solution of (6.3) with a shock starting from* (t_0, x_0).

To simplify the computation of the proof of Theorem 6.1 we assume

$$y_0 = g(y_0) = 0,\ g'(y_0) = -1,\ g^{(3)}(y_0) = 6. \tag{6.13}$$

Then the condition Eq. (6.10) is $yg''(y) > 0$, and Eq. (6.13) gives $t_0 = -1, x_0 = 0$. Denoting $\tau = t - 1$, then we have to construct the solution in a neighborhood $(\tau, x) = (0, 0)$. Next we will extend the function

$$F(t, x) = \frac{f(u_0(y_+(t,x))) - f(u_0(y_-(t,x)))}{u_0(y_+(t,x)) - u_0(y_-(t,x))},$$

and then prove there is a solution of $\frac{d\phi}{dt} = F(t, \phi(t))$ satisfying $\phi(t_0) = 0$ and the curve $x = \phi(t)$ locates in G. The conclusion is obtained by several lemmas.

Lemma 6.1. $\eta_\pm$ *is* C^{p-3} *function of* $\sqrt{\tau}$. *In a neighborhood of* $\tau = 0$ *there are estimates*

$$\eta_\pm \sim \pm\sqrt{\frac{\tau}{3}},\quad x_\pm \sim \mp\frac{2}{\sqrt{3}}\tau^{\frac{3}{2}}. \tag{6.14}$$

Proof. It is enough to study η_+, x_+. When $t > t_0$, $\eta_+ \in C^{p-1}$ is obvious. To discuss its differentiability at $\tau = 0$, we notice $g'(\eta) - g'(0) = 1 - \frac{1}{t}$. By using $g''(0) = 0$ we have

$$\eta^2 \int_0^1 g^{(3)}(\theta\eta)(1 - \theta)d\theta = 1 - \frac{1}{t}.$$

Denote

$$h(\eta, \tau) = \eta \left(\int_0^1 g^{(3)}(\theta\eta)(1 - \theta)d\theta \right)^{\frac{1}{2}} - \sqrt{\frac{\tau}{1 + \tau}},$$

then $\eta_+(t)$ satisfies $h(\eta,\tau)=0$. In view of $h(0,0)=0, \partial_\eta h(0,0)=\sqrt{3}$, we know $\eta\in C^{p-3}$ from the implicit function theorem, and obtain the first estimate in (6.14). Moreover, by virtue of

$$x_+(t)=\eta_+(t)+tg(\eta_+(t))=t(g(\eta_+(t))-g'(\eta_+(t))\eta_+(t)),$$

and the estimate in the neighborhood of $\tau=0$

$$\begin{aligned} g(\eta_+(t))-g'(\eta_+(t))\eta_+(t) &= -\frac{1}{3}g^{(3)}(0)(\eta_+(t))^3+O(\tau^2) \\ &= -2(\eta_+(t))^2+O(\tau^2), \end{aligned}$$

we obtain the estimate on $x_+(t)$. □

It should be noticed that $1+tg'(y)<0$ in the interval $(\eta_-(t),\eta_+(t))$, then $y_\pm(t,x)$, as the function determined by

$$F(t,x,y)\equiv y+tg(y)-x=0,$$

satisfies

$$\frac{dy}{dx}=-\frac{F_x}{F_y}=-\frac{1}{F_y}>0,$$

then $y_\pm(t,x)$ is a monotone increasing function of x. Let

$$\tilde{y}_-(t,x)=\begin{cases} y_-(t,x_+(t)), & \text{if } x\le x_+(t), \\ y_-(t,x), & \text{if } x_+(t)\le x\le x_-(t), \\ \eta_-(t), & \text{if } x\ge x_-(t), \end{cases}$$

$$\tilde{y}_+(t,x)=\begin{cases} y_+(t,x_-(t)), & \text{if } x\ge x_-(t), \\ y_+(t,x), & \text{if } x_+(t)\le x\le x_-(t), \\ \eta_+(t), & \text{if } x\le x_+(t), \end{cases}$$

then $\tilde{y}_\pm$ is a continuous function defined in $t\ge t_0$, satisfying $\tilde{y}_\pm=y_\pm$ in G.

Lemma 6.2. *For $\tilde{y}_\pm(t,x)$ the following estimates hold: (1)*

$$\tilde{y}_-(t,x)=O(\sqrt{\tau}),\ \tilde{y}_+(t,x)=O(\sqrt{\tau}). \tag{6.15}$$

(2) There is $C>0$, such that

$$x(\tilde{y}_-(t,x)+\tilde{y}_+(t,x))\ge -C_\tau|x|. \tag{6.16}$$

Proof. In view of

$$0 \le \eta_+(t) \le \tilde{y}_+(t,x) \le y_+(t,x_-(t)),$$

then the estimate Eq. (6.15) holds, provided $z \equiv y_+(t,x_-(t)) = O(\sqrt{\tau})$. Notice that z satisfies

$$z + tg(z) = x_-(t) = \eta_-(t) + tg(\eta_-(t)),$$
$$1 + tg'(\eta_-(t)) = 0,$$

then

$$g(z) - g(\eta_-(t)) = (z - \eta_-(t))g'(\eta_-(t)).$$

By using Taylor's expansion we have

$$g''(\eta_-(t)) + (z - \eta_-(t)) \int_0^1 (1-\theta)^2 g^{(3)}(\eta_-(t) + \theta(z - \eta_-(t)))d\theta = 0,$$

where $g''(\eta_-(t)) = 6\eta_-(t) + O(\tau)$. Meanwhile,

$$\int_0^1 (1-\theta)^2 g^{(3)}(\eta_-(t) + \theta(z - \eta_-(t)))d\theta \sim 2.$$

Then $z - \eta_-(t) \sim 3\eta_-(t)$, which leads to Eq. (6.15).

To prove Eq. (6.16) we only have to prove

$$\tilde{y}_+(t,0) + \tilde{y}_-(t,0) = O(\tau), \tag{6.17}$$

because $\tilde{y}_+(t,x) + \tilde{y}_-(t,x)$ is a monotone increase function of x. Denote $\xi_\pm(t) = \tilde{y}_\pm(t,0)$, it satisfies

$$0 = \xi + tg(\xi) = \xi(\tau g'(0) + \frac{t}{2}\xi^2 \int_0^1 (1-\theta)^2 g^{(3)}(\theta\xi)d\theta),$$

then

$$\xi_\pm = \pm\sqrt{\frac{2\tau}{t\int_0^1 (1-\theta)^2 g^{(3)}(\theta\xi_\pm)d\theta}}.$$

Therefore, $\xi_\pm \sim \pm\sqrt{\tau}$, and $\xi_+(t) + \xi_-(t) = O(\tau)$ □

Denote $a(x,y) = \dfrac{f(u_0(x)) - f(u_0(y))}{u_0(x) - u_0(y)}$, the problem (6.11) can also be written as

$$\begin{cases} \phi'(t) = a(y_+(t,\phi(t)), y_-(t,\phi(t))), \\ \phi(t_0) = x_0. \end{cases} \tag{6.18}$$

Lemma 6.3. *Assume that $a(x,y)$ is a bounded C^p function, monotone decreasing with respect to x, and satisfies*

$$a(x,y) = -\frac{1}{2}(x+y) + b(x,y), \tag{6.19}$$

where $b(x,y) = O(x^2 + y^2)$.

Proof. Since $a(x, y)$ is symmetric with respect to x and y, and

$$a(y, y) = f'(u_0(y)) = g(y) = -y + O(y^2),$$

then $a(x, y)$ can be written as $a(x, y) = -\frac{1}{2}(x + y) + b(x, y)$ with $b(x, y) = O(x^2 + y^2)$. Moreover, writing $a(x, y)$ as

$$a(x, y) = \int_0^1 f'(u_0(x) + \theta(u_0(y) - u_0(x)))d\theta,$$

and differentiating with respect to x, we obtain

$$\partial_x a(x, y) = \int_0^1 (1 - \theta)u_0'(x)f''(u_0(x) + \theta(u_0(y) - u_0(x)))d\theta.$$

By virtue of $f'' > 0, u_0' < 0$, we have $\partial_x a(x, y) < 0$, then $a(x, y)$ is monotone decreasing with respect to x. □

Lemma 6.4. *Equation (6.18) admits a unique solution* $\phi(t) \in C^1[t_0, +\infty) \cap C^p(t_0, +\infty)$, *satisfying*

$$\phi(t) = O((t - t_0)^2)$$

and

$$x_+(t) < \phi(t) < x_-(t) \quad \text{if} \quad t > t_0. \tag{6.20}$$

Proof. Consider the problem

$$\begin{cases} \phi'(t) = a(\tilde{y}_+(t, \phi(t))), \tilde{y}_-(t, \phi(t))), \\ \phi(t_0) = x_0. \end{cases} \tag{6.21}$$

Since the function $(t, x) \mapsto a(\tilde{y}_+(t, x), \tilde{y}_-(t, \phi(t))$ is continuous and bounded, then according to Piano's theorem the problem admits a solution in $C^1[t_0, +\infty)$. Moreover,

$$\begin{aligned} \frac{d}{dt}(\phi(t)^2) &= 2\phi(t)\phi'(t) \\ &= 2\phi(t)a(\tilde{y}_+(t, \phi(t)), \tilde{y}_-(t, \phi(t))). \end{aligned} \tag{6.22}$$

By using Eq. (6.19) it can be rewritten as

$$\frac{d}{dt}(\phi(t)^2) = -\phi(t)(\tilde{y}_+(t, \phi(t)) + \tilde{y}_-(t, \phi(t))) + O(\tau)\phi(t). \tag{6.23}$$

According to Lemma 6.2 we know

$$\frac{d}{dt}(\phi(t)^2) \leq C\tau|\phi|,$$

and then by integrating

$$|\phi(t)| \le \frac{C}{4}\tau^2.$$

Therefore, by using the estimate for $x_\pm(t)$ in Lemma 6.1 we know that there is $t_1 > t_0$, such that

$$x_+(t) < \phi(t) < x_-(t)$$

holds in (t_0, t_1). Besides, by using the monotonicity of $a(x,y)$ with respect to x, we have

$$\begin{aligned} a(\tilde{y}_+(t, x_-(t)), \tilde{y}_-(t, x_-(t))) &= a(\tilde{y}_+(t, x_-(t)), \eta_-(t)) \\ &< a(\eta_-(t), \eta_-(t)) = g(\eta_-(t)) = x'(t). \end{aligned}$$

Similarly,

$$x'(t) < a(\tilde{y}_+(t, x_+(t)), \tilde{y}_-(t, x_-(t))).$$

Therefore, $x = \phi(t)$ will never intersect with $x = x_+(t)$, so that Eq. (6.21) holds. Furthermore,

$$a(\tilde{y}_+(t, \phi(t)), \tilde{y}_-(t, \phi(t))) = a(\tilde{y}_+(t, \phi(t)), \tilde{y}_-(t, \phi(t))).$$

Then $\phi(t)$ also satisfies Eq. (6.18). Since $y_\pm$ is C^p smooth, then $\phi(t)$ is also C^p smooth, so that the solution of (6.18) is unique.

As we mentioned above, when the solution $\phi(t)$ in G is obtained, the weak solution of Eq. (6.3) can be defined by using Eq. (6.12). Hence Theorem 6.1 is also proved. □

6.1.3 *Estimates of the solution in the neighborhood of the starting point of shock*

Next we are going to establish precise estimates in the neighborhood of the starting point (t_0, x_0) of the shock. This is also the basis of our further discussion.

First let us discuss the estimates of $y_\pm(t,x)$ in the neighborhood of (t_0, x_0). Let

$$s = \sqrt{\tau},\ \lambda = \frac{x}{s^3},\ \mu = \frac{y}{s},\ v = (\tau^3 + x^2)^{1/3}, \tag{6.24}$$

we have the following lemma.

Lemma 6.5.

$$y_\pm = s\left(\pm 1 + \frac{\lambda}{2} - \frac{g^{(4)}(0)}{48}s\right) + O(s^3 + s\lambda^2), \tag{6.25}$$

$$\frac{2}{3}v \le y_\pm \le \frac{3}{2}v. \tag{6.26}$$

Proof. Both $y_\pm(t,x)$ satisfy Eq. (6.6), by denoting

$$h(y) = \int_0^1 \frac{(1-\theta)^2}{2} g^{(3)}(\theta y)d\theta - 1,$$

then $h \in C^{p-3}, h(0) = 0$. By using the property of the derivatives of $g(y)$, we have

$$g(y) = -y + y^3(h(y)+1),$$

then Eq. (6.6) becomes

$$y + t(-y + y^3 + h(y)y^3) = x.$$

Substituting Eq. (6.24) into it gives

$$F(s,\lambda,\mu) \equiv (1+s^2)\mu^3 - \mu + (1+s^2)\mu^3 h(s\mu) - \lambda = 0. \tag{6.27}$$

There are three roots of Eq. (6.27) as $(s,\lambda) = (0,0)$. Since $\partial_\mu F(0,0,\mu_\pm) = 2$, then the implicit theorem implies $\mu_\pm$ can be solved as a C^{p-1} function of s,λ. Therefore, one can write $\mu_\pm$ as the form $\pm 1 + as + b\lambda + O(s^2+\lambda^2)$ with

$$a = \mu_s = -\frac{F_s}{F_\mu} = -\frac{1}{2}\mu^4 h'(0) = \frac{1}{48}g^{(4)}(0),$$

$$b = \mu_\lambda = -\frac{F_\lambda}{F_\mu} = \frac{1}{2}.$$

Then Eq. (6.25) is obtained.

To prove Eq. (6.26), we consider the estimate of y_+. Denote

$$T = \frac{\tau}{v^2},\ X = \frac{x}{v^3},\ Y = \frac{y}{v},$$

Eq. (6.27) can be written as

$$F(\tau,T,X,Y) \equiv X + TY - (1+\tau)Y^3(h(vY)+1) = 0. \tag{6.28}$$

Here we have to imply that $Y \in \left(\frac{2}{3},\frac{3}{2}\right)$, and it is enough to imply that $Y \in (\frac{3}{4},\frac{4}{3})$ for $v=0$. Notice that Eq. (6.28) can be reduced to

$$F_0(Y) \equiv Y^3 - TY - X = 0 \tag{6.29}$$

for $v = 0$. Obviously, $T^3 + X^2 = 1$ implies $T + X \geq 1$. Notice that

$$F_0(0) \leq 0,\quad F_0'(0) + F_0(0) = -T - X < 0,$$

then F_0 is negative for small Y, so that $F_0(Y) = 0$ has at least one positive root. In view of $F_0(Y) < 0$ for small positive Y, then the number of positive roots of $F_0(Y) = 0$ must be 3, if it is larger than 1. But in this case the number of positive roots of $F_0'(Y) = 0$ will be 2. This is impossible, because $F_0' = 3Y^2 - T$, whose roots must be one positive and one negative. The contradiction indicates that the positive root of $F_0(Y) = 0$ is unique.

Based on the uniqueness of the positive root, one can estimate the bound of the positive root Y_0. Since

$$F_0\left(\frac{4}{3}\right) = \frac{64}{27} - \frac{4}{3}T - X > \frac{64}{27} - \frac{4}{3} - 1 > 0,$$

then $Y_0 < \frac{4}{3}$. On the other hand,

$$F_0\left(\frac{3}{4}\right) = \frac{27}{64} - \frac{3}{4}T - X < \frac{27}{64} - \frac{3}{4}(T + X) < 0,$$

then $Y_0 > \frac{3}{4}$. Combining the above discussions, we know that for small positive τ the root of Eq. (6.28) locates in $\left(\frac{2}{3}, \frac{3}{2}\right)$. Hence (6.25) holds. □

By using above lemmas we can prove the following estimates of the weak solution near the starting point of shock.

Theorem 6.2. *In the neighborhood of (t_0, x_0), the weak solution $u(t, x)$ satisfies*

$$\begin{cases} |u(t,x) - u(t_0,x_0)| \le C_0((t-t_0)^3 + (x-x_0)^2)^{1/6}, \\ |\partial_t u(t,x)| \le C_0((t-t_0)^3 + (x-x_0)^2)^{-1/6}, \\ |\partial_x u(t,x)| \le C_0((t-t_0)^3 + (x-x_0)^2)^{-1/3}, \\ |\partial_{xx} u(t,x)| \le C_0((t-t_0)^3 + (x-x_0)^2)^{-5/6}. \end{cases} \tag{6.30}$$

Proof. In the neighborhood of (t_0, x_0) the solution $u(t, x)$ is defined by Eq. (6.12). When $x > \phi(t)$,

$$|u(t,x) - u(t_0,x_0)| = |u_0(y_+(t,x)) - u_0(y_+(t_0,x_0))| \le C_0|y_+(t,x)|,$$

then Eq. (6.27) implies Eq. (6.30). In order to obtain estimates of the derivatives of the solution, we derive estimates for the derivatives of $y_+(t, x)$. Obviously,

$$\partial_x y_+(t,x) = \frac{1}{1 + tg'(y_+(t,x))}. \tag{6.31}$$

According to the assumption on $g(y)$, we have

$$\begin{aligned} 1 + tg'(y_+(t,x)) &= 1 + t(-1 + 3y_+^2(t,x)) + O((\tau^3 + x^2)^{1/2}) \\ &= 3y_+^2(t,x) - \tau + O((\tau^3 + x^2)^{1/2}) \ge \frac{1}{2}(3y_+^2(t,x) - \tau). \end{aligned} \tag{6.32}$$

For sufficiently small τ,

$$3y_+^2(t,x) - \tau \geq \frac{4}{3}(\tau^3 + x^2)^{1/3} - \tau \geq c(\tau^3 + x^2)^{1/2},$$

Hence

$$|\partial_x(u_0(y_+(t,x))| \leq |u_0'| \cdot |\partial_x y_+(t,x)| \leq C_0(\tau^3 + x^2)^{-1/3}. \tag{6.33}$$

Similarly, by using

$$\partial_t y_+(t,x) = -\frac{g(y_+(t,x))}{1 + tg'(y_+(t,x))} \tag{6.34}$$

and Eq. (6.25) we know

$$|g(y_+(t,x))| \leq C|y_+(t,x)| \leq C\sqrt{\tau},$$

and obtain the third estimate in (6.30). Finally,

$$\begin{aligned}
|\partial_{xx} u_0(y_+(t,x))| &\leq |u_0'' \cdot (\partial_x y_+(t,x))^2| + |u_0' \cdot (\partial_{xx} y_+(t,x))| \\
&\leq C\left((\tau^3 + x^2)^{-1/3} + |\frac{tg''\partial_x y_+}{(1+tg')^2}|\right).
\end{aligned}$$

For the second term in the right hand side one can use $g''(0) = 0$ and (6.31), (6.32) to obtain

$$|\frac{tg''\partial_x y_+}{(1+tg')^2}| \leq C(\tau^3 + x^2)^{-5/6}.$$

Hence the forth estimate of Eq. (6.30) is obtained. □

6.2 The case of system

The formation of shocks for systems of conservation laws is more complicated than that for scalar equation. Next we are going to discuss the formation of shocks for Euler system in gas dynamics. We will mainly discuss the planar isentropic and irrotational flow, which ia a 2×2 system. From the discussion here readers can figure out the new difficulty arisen in the system case. Since near the point where shock forms the strength of the shock is weak, then the equation of isentropic and irrotational flow is a good approximation of the Euler equation. In the last section we give a conclusion on formation of shocks for the full Euler system. The result gives a more precise description of shock formation in the real gas dynamics.

6.2.1 Background and conclusion

The system describing the planar irrotational flow is

$$\begin{cases}(a^2-u^2)u_x - uv(u_y+v_x)+(a^2-v^2)v_y=0,\\ u_y=v_x,\end{cases} \tag{6.35}$$

where (u,v) is velocity, a is sonic speed, which equals $(\gamma\rho^{\gamma-1})^{1/2}$ with γ being the adiabatic exponent. a,u,v satisfies a Bernoulli's relation

$$\frac{1}{2}(u^2+v^2)+\frac{a^2}{\gamma-1}=const. \tag{6.36}$$

Assume that there is a uniform flow coming from infinity. The direction of the velocity is parallel to x-axis, which is considered as a rigid wall as $x<0$. The velocity (u,v) is assumed as supersonic, i.e. $(u,v)=(q_0,0)$ with $q_0>a_0$. Assume that the flow locates above the rigid wall, which is bending up starting from the origin. Then the gas is compressed and a shock will be formed due to the compression.

The Rankine-Hugoniot conditions on shocks for (6.35) is

$$[u]+\sigma[v]=0,\quad \sigma[\rho u]-[\sigma v]=0. \tag{6.37}$$

Then any state (u,v,ρ), which can be connected with a given state (u_0,v_0,ρ_0), has to satisfy

$$(\rho u-\rho_0u_0)(u-u_0)+(\rho v-\rho_0v_0)(v-v_0)=0. \tag{6.38}$$

Denote $\theta=\arctan(v/u)$, $q=\sqrt{u^2+v^2}$, Eq. (6.38) becomes

$$\cos(\theta-\theta_0)=\frac{\rho q^2+\rho_0q_0^2}{(\rho+\rho_0)q_0q}. \tag{6.39}$$

Equation (6.38) or Eq. (6.39) is called **shock polar equation**.

Let us first construct the compressible simple waves by using the characteristic method. The characteristic direction of Eq. (6.35) is

$$\lambda_\pm=\frac{uv\pm a\sqrt{u^2+v^2-a^2}}{u^2-a^2}. \tag{6.40}$$

The Riemann invariants of Eq. (6.35) is

$$r=\theta+F(q),\quad s=\theta-F(q), \tag{6.41}$$

where $F(q)=\displaystyle\int\frac{\sqrt{q^2-a^2}}{aq}dq$. By using the Riemann invariants Eq. (6.35) can be reduced to

$$\begin{cases}\dfrac{\partial s}{\partial x}+\lambda_+\dfrac{\partial s}{\partial y}=0,\\[2mm] \dfrac{\partial r}{\partial x}+\lambda_-\dfrac{\partial r}{\partial y}=0.\end{cases} \tag{6.42}$$

Hence r (s resp.) is a constant on λ_- (λ_+ resp.) characteristics. If the flow comes from left and locates above the wall, then from any point in the domain, where the smooth solution exists, there is a characteristics of first class comes from $x < 0$. Therefore, $r \equiv$ *const.* in the whole domain above the wall, and all second characteristics of second class are straight lines. Meanwhile, (u, v) are constants on each characteristics of second class. Such a solution is called **simple wave of second class**. The image of the whole domain of the simple wave under the map

$$T : (x, y) \mapsto (u, v) = (u(x, y), v(x, y))$$

is an epicycloidal on (u, v) plane

$$\begin{cases} u = a_*(\cos\mu(\omega - \omega_*)\cos\omega + \mu^{-1}\sin\mu(\omega - \omega_*)\sin\omega), \\ v = a_*(\cos\mu(\omega - \omega_*)\sin\omega - \mu^{-1}\sin\mu(\omega - \omega_*)\cos\omega), \end{cases} \tag{6.43}$$

where $\mu = (\frac{\gamma - 1}{\gamma + 1})^{\frac{1}{2}}$, a_* is the critical sonic speed, both a_* and ω_* are determined by the incoming flow.

Assume that the equation of the wall is $y = f(x)$, which is 0 for $x < 0$, and satisfies $f'(x) > 0, f''(x) > 0$ for $x > 0$. This means that the wall is convex downward. Since the direction of the velocity of the flow should be parallel to the wall at the surface of the wall, then for any $\bar{x} \geq 0$, the direction of the velocity at $(\bar{x}, f(\bar{x}))$ is $u : v = 1 : f'(\bar{x})$, hence we have

$$f(\bar{x}) = \frac{\cos\mu(\omega - \omega_*)\sin\omega - \mu^{-1}\sin\mu(\omega - \omega_*)\cos\omega}{\cos\mu(\omega - \omega_*)\cos\omega + \mu^{-1}\sin\mu(\omega - \omega_*)\sin\omega}.$$

Direct computations indicate that the derivative of the right hand side with respect to ω is positive, so that ω is a monotone increasing function of $\bar{x}$ due to $f''(x) > 0$. On (x, y) plane the characteristics corresponding to the parameter ω is

$$\begin{cases} x = \bar{x} + t\cos(\omega - \pi/2) \ (= \bar{x} + t\sin\omega), \\ y = f(\bar{x}) + t\sin(\omega - \pi/2) \ (= f(\bar{x}) + t\cos\omega). \end{cases} \tag{6.44}$$

Obviously, if $\bar{x}_1 > \bar{x}_2$, then the slope of the corresponding characteristics satisfies $\tan(\omega_1 - \pi/2) > \tan(\omega_2 - \pi/2)$. Hence these two characteristics must intersect at some point P on the upper half plane. Hence the domain of the smooth solution cannot be extended to the point P, so that the solution must blow up before arriving P.

Like the discussion in Section 6.1 we can determine the starting point of the shock by making the envelop of characteristics. The envelop of the family (6.44) is

$$\Delta \equiv \left| \begin{pmatrix} x_{\bar{x}} & x_t \\ y_{\bar{x}} & y_t \end{pmatrix} \right| = 0 \tag{6.45}$$

Substituting Eq. (6.44) into it gives

$$\Delta \equiv \cos\omega - t\omega' + f'\sin\omega.$$

Therefore, on the envelop we have $t = \dfrac{1}{\omega'}(f'\sin\omega + \cos\omega)$, and the equation of the envelop with parameter $\bar{x}$ is

$$\begin{cases} x = \bar{x} - \dfrac{1}{\omega'}(f'\sin\omega + \cos\omega)\sin\omega, \\ y = f(\bar{x}) + \dfrac{1}{\omega'}(f'\sin\omega + \cos\omega)\cos\omega. \end{cases} \tag{6.46}$$

Substituting ω, ω' as functions of $\bar{x}$ into the first equation of Eq. (6.46) and denoting the result by $h(\bar{x})$, we have

$$h'(\bar{x}) = 0, \;\; h''(\bar{x}) > 0$$

at some point $\bar{x} = \bar{x}_0$. This means that the coordinate of the points on the envelop takes minimum at $\bar{x}_0$. The fact implies that $(\bar{x}_0, \bar{y}_0)$ is the cusp of the envelop.

The main conclusion in this section is

Theorem 6.3. *Assume that $f'(x)$ satisfies the above assumptions, so that the family of characteristics of second class of the system (6.35) forms an envelop with cusp (x_0, y_0), then in a neighborhood Ω of (x_0, y_0) there is an entropy solution of (6.35), which contains a shock $\Gamma: \;\; y = \phi(x)$ starting from (x_0, y_0). The solution (u, v) is continuous on $\Omega \setminus \Gamma$, and satisfies the following estimates*

$$\begin{cases} \phi(x) = y_0 + \alpha(x - x_0) + O((x - x_0)^2), \\ r(x, y) = O((x - x_0)^{3/2}), \\ s(x, y) = s(x_0, y_0) + O((x - x_0)^3 + (y - y_0 - \alpha(x - x_0)^2)^{1/6}), \end{cases} \tag{6.47}$$

where α is the slope of the characteristics of second class passing through the cusp (x_0, y_0).

The outline of the proof is as follows. As mentioned above, the solution in $x < x_0$ is a simple wave, in which (u, v) takes constant on each characteristics (6.46) determined by Eq. (6.42). For $x > x_0$, the solution is singular, which will be constructed by an iterative scheme. Let $r^{(0)}(x, y)$ be a constant, substituting it into Eq. (6.41) gives an equation for s. By solving s we obtain $s^{(0)}(x, y)$, and then obtain the location of an approximate shock by using $r^{(0)}$ and $s^{(0)}$. Successively determine approximate solutions in both sides of the approximate shock as update the location of the approximate shock, we can establish a sequence $\{r^{(\nu)}(x, y)\}$ and $\{s^{(\nu)}(x, y)\}$. Then the limit of the sequence gives the required solution with a shock starting from (x_0, y_0). Next we will prove that the sequence of approximate solutions can be well defined and the sequence is convergent.

6.2.2 *The property of the first approximate solution*

Since our discussion is proceeded in a neighborhood of the starting point (x_0, y_0) of shock, we may assume that the characteristics of second class in $\{x = 0, y \geq 0\}$ cover all neighborhood of (x_0, y_0). Hence we only need to consider an initial value problem of (6.42). Besides, to simplify the computations we can take the starting point as the origin, and take the second characteristics passing the starting point as x-axis. Moreover, the initial line is placed on $x = -1$, the Riemann invariant r is assumed to be 0 as $x < 0$.

The first step in to construct $s^{(0)}(x, y)$. In its smooth domain, $s^{(0)}(x, y)$ satisfies

$$\partial_x s(x, y) + \lambda_+ \partial_y s(x, y) = 0, \tag{6.48}$$

$$s(-1, y) = s_0(y). \tag{6.49}$$

Since the weak solution depends on the form of the conservation laws corresponding to Eq. (6.48), and different form of conservation laws leads to different Rankine-Hugoniot conditions, then we have to choose a suitable one, which coincide with Eq. (6.37). Denote the inverse function of $F(q)$ by G, we have

$$q = G\left(\frac{r - s}{2}\right), \quad G'(F(q)) = aq(q^2 - a^2)^{-1/2}.$$

Moreover, let

$$\begin{aligned} e(u, v) &= u - av(q^2 - a^2)^{-1/2} \\ &= q\cos\theta - aq\sin\theta(q^2 - a^2)^{-1/2} \\ &= 2G\left(\frac{r - s}{2}\right)\sin\frac{r + s}{2}, \end{aligned}$$

then

$$\begin{aligned} e\lambda_+ &= v = au(q^2 - a^2)^{-1/2} \\ &= aq\cos\theta(q^2 - a^2)^{-1/2} + q\sin\theta \\ &= -2G\left(\frac{r - s}{2}\right)\cos\frac{r + s}{2}. \end{aligned}$$

Multiplying Eq. (6.48) by e and taking $r = 0$ we obtain

$$\left(G\left(-\frac{s}{2}\right)\sin\frac{s}{2}\right)_x + \left(-G\left(-\frac{s}{2}\right)\cos\frac{s}{2}\right)_y = 0, \tag{6.50}$$

which is the form of conservation law to look for $s^{(0)}(x, y)$. In accordance, the shock condition is

$$\sigma = -\frac{[2G(-s/2)\cos s/2]}{[2G(-s/2)\sin s/2]}. \tag{6.51}$$

Obviously, the numerator and the denominator are $[u]$ and $[v]$ respectively, as $r=0$.

The property of the solution $s^{(0)}(x,y)$ of (6.48), (6.49) can be obtained by using the result in Section 6.1. To have it we denote

$$m = G(-s/2)\sin(s/2),\ f(m) = G(-s/2)\cos(s/2).$$

Since the inverse $s(m)$ of $m(s)$ exists, then $f(m) = [G(-s/2)\cos(s/2)]_{s=h(m)}$ is also well defined. Hence (6.48), (6.49) can be written as

$$\begin{cases} m_x + (f(m))_y = 0, \\ m(-1,y) = m(s_0(y)). \end{cases} \tag{6.52}$$

By applying Theorems 6.1 and 6.2 we obtain the existence and estimate of $m(x,y)$ and $y=\phi(x)$. In accordance, the estimates Eq. (6.30) hold for $s^{(0)}(x,y)$.

To avoid the trouble caused by the change of the location of approximate shocks, we will fix it onto the x-axis by a coordinate transformation. For instance, if the approximate shock is $y=\phi^{(\nu)}(x)$ for $x\geq 0$ in ν^{-th} step, then the transformation is

$$E^{(\nu)}: x_1 = x,\ y_1 = \begin{cases} y - \phi^{(\nu)}(x), & \text{if } x \geq 0, \\ y, & \text{if } -1 \leq x \leq 0. \end{cases} \tag{6.53}$$

Denote

$$\sigma^{(\nu)} = \begin{cases} (\phi^{(\nu)})', & \text{if } x \geq 0, \\ 0, & \text{if } -1 \leq x \leq 0. \end{cases}$$

Again denote the new coordinates by (x,y), then $s^{(0)}(x,y)$ satisfies

$$\begin{cases} \partial_x s^{(0)}(x,y) + (\lambda_+ - \sigma^{(0)})\partial_y s^{(0)}(x,y) = 0, \\ s^{(0)}(-1,y) = s_0(y). \end{cases} \tag{6.54}$$

Denote by $y=\eta_\pm(x,a,b)$ the characteristics passing from (a,b) corresponding to $\pm b>0$. Denote $\eta_0=\eta(-1,a,b)$, and denote the function $y_\pm(t,x)$ in the previous section by $z_\pm(a,b)$. Let $g(z)=\lambda_+(s_0(z))$, then we have $g(0)=g''(0)=0$ and $g'(0)=-1$. Besides, to simplify the computations we also assume $g^{(3)}(0)=-6$.

Lemma 6.6. *In the neighborhood of the origin the characteristics of Eq. (6.54) will not intersect with x-axis. Moreover, the following estimates for the points on the characteristics hold*

$$\pm\eta_\pm(0,a,b) = \sqrt{a} + O(a) + O_1(|b|^{1/3}), \tag{6.55}$$

$$\pm(\eta_\pm(x,a,b) - b) = \sqrt{a}(a-x) + O_1(|b|^{1/3}(a-x)) + O(a(a-x)), \tag{6.56}$$

where $O_1(|b|^{1/3})$ means a positive number no more than $O(|b|^{1/3})$.

Proof. Let us only consider the case $b > 0$ and denote η_+ by η, then

$$\eta(x,a,b) + \phi(x) = \eta_0 + (x+1)g(\eta_0). \tag{6.57}$$

Taking $x = a$ gives

$$b + \phi(a) = \eta_0 + (a+1)g(\eta_0),$$

then

$$b - \eta(x,a,b) = \phi(x) - \phi(a) + (a-x)g(\eta_0).$$

$\phi(x) = O(x^2)$ implies $\phi(x) - \phi(a) = O(a(a-x))$, then

$$b - \eta(x,a,b) = -\eta_0(a-x) + O(\eta_0^3(a-x) + a(a-x)). \tag{6.58}$$

Notice

$$\eta_0 = z_+(a,b) = z_+(a,0) + b\partial_b z_+(a,\theta b), \quad \theta \in (0,1),$$

and apply the estimates (6.25), (6.31), we have

$$\eta_0 = a^{1/2} + b^{1/3}b^{2/3}(a^3+b^2)^{-1/3} + O(a) = a^{1/2} + cb^{1/3} + O(a), \tag{6.59}$$

where c is a bounded positive number. This is the estimate (6.55). Substituting it into Eq. (6.58) leads to Eq. (6.56). □

Lemma 6.7. *Denote* $I = \int_0^a (-\lambda_{+s} \cdot s_y^{(0)})(x, \eta(x,a,b))dx$, *the integrand is positive, and the integral satisfies*

$$|I| \le \log\frac{3}{2} + C\sqrt{a}. \tag{6.60}$$

Proof. $s^{(0)}$ is constant on characteristics of second class, then

$$s^{(0)}(x,\eta(x,a,b)) = s^{(0)}(-1,\eta_0) = s_0(\eta_0),$$

$$\partial_y s^{(0)}(x,\eta(x,a,b)) \cdot \eta_0 = s_0'(\eta_0) \cdot \eta_{0b}.$$

Equation (6.58) implies $\eta_b = \eta_{0b}(1 + (x+1)g'(\eta_0))$, then

$$\partial_y s^{(0)}(x,\eta(x,a,b)) = \frac{s_0'(\eta_0)}{1+(x+1)g'(\eta_0)},$$

$$\begin{aligned}\lambda_{+s} \cdot s_y^{(0)} &= \partial_b(\lambda_+(s^{(0)}(x,\eta(x,a,b)))) \cdot \eta_b^{-1} \\ &= g'(\eta_0) \cdot \eta_{0b} \cdot \eta_b^{-1} = \frac{g'(\eta_0)}{1+(x+1)g'(\eta_0)}.\end{aligned}$$

In view of the property of g', we know the fraction is negative identically. This is the first conclusion of the lemma.

Furthermore,

$$|I| = -\int_0^a \frac{g'(\eta_0)}{1+(1+x)g'(\eta_0)}dx$$
$$= -\log\frac{1+(1+a)g'(\eta_0)}{1+g'(\eta_0)}.$$

Denote $\tilde{\eta}_0 = \eta(-1,a,0)$, we have $\eta_0 > \tilde{\eta}_0$ for $b > 0$, then $g'(\eta_0) \geq g'(\tilde{\eta}_0)$ and

$$-\log\frac{1+(1+a)g'(\eta_0)}{1+g'(\eta_0)} \leq -\log\frac{1+(1+a)g'(\tilde{\eta}_0)}{1+g'(\tilde{\eta}_0)}.$$

Since $\tilde{\eta}_0 = \sqrt{a}+O(a)$, then

$$1+g'(\tilde{\eta}_0) = 1+\left(-1+\frac{1}{2}g^{(3)}(0)\tilde{\eta}_0^2+O(a^{3/2})\right) = 3a+O(a^{3/2}),$$

$$1+(a+1)g'(\tilde{\eta}_0) = 2a+O(a^{3/2}),$$

$$-\log\frac{1+(1+a)g'(\tilde{\eta}_0)}{1+g'(\tilde{\eta}_0)} = -\log\left(\frac{2}{3}(1+O(a^{1/2}))\right).$$

Hence we obtain $I \leq \log\frac{3}{2}+C\sqrt{a}$. □

Remark If $\zeta(x,a,b)$ is a function satisfying

$$|\zeta(x,a,b)-\eta(x,a,b)| \leq Ca(a-x), \tag{6.61}$$

then the following estimate

$$\left|\int_0^a (\lambda_{+s}\cdot s_y^{(0)})(x,\zeta(x,a,b))dx\right| \leq \log\frac{3}{2}+C\sqrt{a}. \tag{6.62}$$

We leave the proof to readers.

Lemma 6.8. *If $a > 0$, $a, |b|$ are sufficiently small, then on the characteristics $y = \eta(x,a,b)$*

$$x^3+y^2 \geq \frac{5}{16}(a^3+b^2). \tag{6.63}$$

Proof. Let us only consider the case $b > 0$. Equation (6.56) implies

$$\eta(x,a,b)-b = \sqrt{a}(a-x)+O(a(a-x))+O_1(b^{\frac{1}{3}}(a-x)). \tag{6.64}$$

When $b = 0$, Eq. (6.64) becomes

$$\eta(x,a,0) \geq \sqrt{a}(a-x)+O(a(a-x)),$$

then there is a constant C, such that

$$\eta(x,a,b) - b \geq \eta(x,a,0) - Ca(a-x).$$

Hence for small a we have

$$\eta(x,a,b) \geq b + \frac{\sqrt{3}}{2}\sqrt{a}(a-x),$$

which implies

$$x^3 + y^2 \geq b^3 + \frac{3}{4}a(a-x)^2 + x^3.$$

Since $\frac{3}{4}a(a-x)^2 + x^3$ takes minimum at $x = \frac{a}{2}$, then

$$x^3 + y^2 \geq b^3 + \frac{5}{16}a^3 \geq \frac{5}{16}(a^3 + b^2).$$

□

The characteristics of first class run across the shock. Then in order to estimate approximate solution $r^{(\nu)}(x,y)$ we have to carefully analyze the relations of jumps of all flow parameters on the shock.

Lemma 6.9. *For θ, q defined as above, we have*

$$[\theta] = k[F(q)], \tag{6.65}$$

where k is a smooth function of $\theta_-, \theta_+, q_-, q_+$, which has limit -1, as $[q] \to 0$.

Proof. Equation (6.52) implies

$$\begin{aligned}
\sin^2\theta &= 1 - \frac{((\rho_+ q_+^2 + \rho_- q_-^2)^2}{(\rho_+ + \rho_-)q_+ q_-)^2} \\
&= -\frac{(q_+^2 - q_-^2)(\rho_+^2(q_+^2 - q_-^2) + q_-^2(\rho_+^2 - \rho_-^2))}{((\rho_+ + \rho_-)q_+ q_-)^2} \\
&= -\frac{(q_+ + q_-)^2(\rho_+^2 + q_-^2(\rho_+^2 - \rho_-^2)/(q_+^2 - q_-^2))}{((\rho_+ + \rho_-)q_+ q_-)^2}\frac{[q]^2}{[F(q)]^2}[F(q)]^2.
\end{aligned}$$

When $[q] \to 0$, the coefficient of $[F(q)]^2$ tends to

$$-\frac{4q^2(\rho^2 + q^2\frac{\rho}{q}\frac{d\rho}{dq})}{4\rho^2 q^2 (F'(q))^2} = -\frac{(\rho^2 + \rho q \frac{d\rho}{dq})}{4\rho^2 q^2(F'(q))^2}.$$

Bernoulli's relation gives

$$qdq + \gamma\rho^{\gamma-2}d\rho = 0.$$

Then $\frac{d\rho}{dq} = -\rho q/a^2$, and

$$-\frac{\rho^2 + \rho q d\rho/dq}{\rho^2 q^2 F'(q)^2} = -\frac{\rho^2 - \rho^2 q^2 a^{-2}}{\rho^2 q^2 (q^2 - a^2) a^{-2} q^{-2}} = 1.$$

Due to $\sin[\theta] \sim [\theta]$, we obtain Eq. (6.65), where $|k|$ has limit 1, as $[q]$ tends to 0. Notice that $[\theta]$ and $[q]$ take different sign, $[q]$ and $[F(q)]$ take same sign, then $[\theta]$ and $[F(q)]$ take different sign. Hence $k \to -1$, as $[q] \to 0$. □

Lemma 6.10. *Assume $r_- = 0$ on the lift side of shock, then*

$$r_+ = f(s_+, s_-)[s]^3, \tag{6.66}$$

where $f(s_+, s_-)$ is a smooth function of its arguments.

Proof. From Lemma 6.9 we have

$$[\theta] = (1 + h(\theta_+, \theta_-, q_+, q_-)[F(q)])[F(q)]. \tag{6.67}$$

Exchange plus sign with minus sign, we have

$$h(\theta_-, \theta_+, q_-, q_+) = -h(\theta_+, \theta_-, q_+, q_-),$$

then h vanishes for $[\theta] = [q] = 0$, so that Eq. (6.67) can be written as

$$[\theta] = (1 + h_1[F(q)]^2)[F(q)]$$

or

$$[\theta]^2 = (1 + h_2[F(q)]^2)[F(q)]^2,$$

where h_1 and h_2 are suitable smooth functions, and vanishes as $[\theta] = [q] = 0$. By substituting the expression of r and s, we have

$$[s + r]^2 = (1 + \frac{h_2}{4}[s - r]^2)[s - r]^2.$$

Hence

$$r_+([s - r] + r_+) = \frac{1}{16} h_2 [s - r]^4 \tag{6.68}$$

due to $r_- = 0$. Denote $z = \frac{r_+}{[s - r]}$, then z satisfies

$$z(1 + z) = \frac{1}{16} h_2 [s - r]^2. \tag{6.69}$$

Lemma 6.9 implies

$$z = -\frac{1}{2}[\theta + F(q)]/[F(q)] \to 0$$

Then the root of Eq. (6.69) is

$$z = -\frac{1}{2} + \frac{1}{2}\sqrt{1 + \frac{1}{4} h_2 [s - r]^2},$$

so that $r_+ = h_3[s - r]^3$, where h_3 is smooth. Now let $w = r_+/[s]^3$, then $[s - r] = [s] - w[s]^3$, and Eq. (6.68) gives

$$w - h_3(w[s]^3, s_+, s_-)(1 - w[s]^2)^3 = 0. \tag{6.70}$$

Obviously, the derivative of the left hand side with respect to w is 1 for $[s] = 0$. Then by using the implicit function theorem w can be solved as a smooth function of s_+ and s_-. □

6.2.3 *Estimates and convergence of the sequence of approximate solutions*

Based on the above preparation we can establish the sequence of approximate solutions $\{r^{(\nu)}(x,y), s^{(\nu)}(x,y)\}$. Let $\Omega = \Omega_0 \cup \Omega_+ \cup \Omega_-$ be the domain of definition of the sequence, where

$$\begin{aligned}\Omega_0 &= \{(x,y);\ -1 \le x \le 0, -\epsilon \le y \le \epsilon\},\\ \Omega_+ &= \{(x,y);\ 0 \le x \le \eta, 0 \le y \le \epsilon - x\},\\ \Omega_- &= \{(x,y);\ 0 \le x \le \eta, -\epsilon + x \le y \le 0\},\end{aligned}$$

and η,ϵ are small numbers.

For the first term of the sequence, $r^{(0)}(x,y) = 0$, $s^{(0)}(x,y)$ is determined by Eq. (6.54). For $\nu \ge 0$, $\sigma^{(\nu)}, r^{(\nu+1)}, s^{(\nu+1)}$ can be defined by induction as follows:

$$\sigma^{(\nu)} = -\frac{[G((r^{(\nu)} - s^{(\nu)})/2)\cos(r^{(\nu)} + s^{(\nu)})/2]}{[G((r^{(\nu)} - s^{(\nu)})/2)\sin(r^{(\nu)} + s^{(\nu)})/2]}. \tag{6.71}$$

$$\begin{cases} \partial_x r^{(\nu+1)}(x,y) + (\lambda_-^{(\nu)} - \sigma^{(\nu)})\partial_y r^{\nu+1)}(x,y) = 0,\\ \partial_x s^{(\nu+1)}(x,y) + (\lambda_+^{(\nu)} - \sigma^{(\nu)})\partial_y s^{\nu+1)}(x,y) = 0,\\ r^{(\nu+1)}(-1,y) = 0,\ \ s^{(\nu+1)}(-1,y) = s_0(y), \end{cases} \tag{6.72}$$

where $\lambda_\pm^{(\nu)} = \lambda_\pm(r^{(\nu)}, s^{(\nu)})$.

We confirm that $\sigma^{(\nu)}, r^{(\nu+1)}, s^{(\nu+1)}$ are well defined by Eq. (6.71) and Eq. (6.72). To indicate it we are going to prove the following propositions.

$$F_1^{(\nu)}: \begin{cases} 1)\ s^{(\nu)} = s^{(0)} \quad \text{in } \Omega_0 \cup \Omega_+,\\ 2)\ s^{(\nu)} \in C^1(\bar{\Omega}_- \setminus (0,0)),\\ 3)\ |s^{(\nu)}(x,y) - s^{(0)}(x,y)| \le Cx,\\ 4)\ |\partial_y(s^{(\nu)} - s^{(0)})(x,y)| \le C(x^3 + y^2)^{-1/6}. \end{cases}$$

$$F_2^{(\nu)}: \begin{cases} 1)\ r^{(\nu)} = 0 \quad \text{in } \Omega_0 \cup \Omega_+,\\ 2)\ r^{(\nu)} \in C^1(\bar{\Omega}_- \setminus (0,0)),\\ 3)\ |r^{(\nu)}(x,y)| \le Cx^{3/2},\\ 4)\ |\partial_y r^{(\nu)}| \le C(x^3 + y^2)^{1/2}. \end{cases}$$

The constant C in the above two propositions is independent of ν. Obviously, the propositions are valid for the index $\nu = 0$, we now indicate that the validity for index ν implies its validity for index $\nu + 1$ by a set of lemmas.

Lemma 6.11. *If $F_{1,2}^{(\nu)}$ holds, then*

$$|\sigma^{(\nu)} - \sigma^{(0)}| \le Cx. \tag{6.73}$$

Proof. Since $r^{(\nu)}, s^{(\nu)}$ are independent of ν in Ω_+, then Eq. (6.71) can be written as

$$\sigma^{(\nu)} = H(r,s)|_{r=r^{(\nu)},s=s^{(\nu)}}, \tag{6.74}$$

where

$$H(r,s) = -\frac{[G((r-s)/2)\cos(r+s)/2]}{[G((r-s)/2)\sin(r+s)/2]}.$$

Hence

$$|\sigma^{(\nu)} - \sigma^{(0)}| \le C(|r^{(\nu)}| + |s^{(\nu)} - s^{(0)}|) \le Cx.$$

□

Denote by $y = \eta^{(\nu)}(x,a,b)$ the characteristics of the second equation in Eq. (6.72) passing through (a,b), then we have the following lwmma.

Lemma 6.12. *If $F_{1,2}^{(\nu)}$ holds, then*

$$|\eta^{(\nu)}(x,a,b) - \eta^{(0)}(x,a,b)| \le Ca(a-x). \tag{6.75}$$

Proof. By virtue of $b = \eta^{(\nu)}(a,a,b)$ we have

$$b - \eta^{(\nu)}(x,a,b) = \int_x^a (\lambda_+^{(\nu)}(\alpha, \eta^{(\nu)}(\alpha,a,b)) - \sigma^{(\nu)}(\alpha))d\alpha. \tag{6.76}$$

To estimate $\eta^{(\nu)}(x,a,b)$ we introduce another iterative process. Temporarily fixed ν, and define

$$\zeta_0(x,a,b) = \eta^{(0)}(x,a,b), \tag{6.77}$$

$$\zeta_{n+1}(x,a,b) = b - \int_x^a (\lambda_+^{(\nu)}(\alpha, \zeta_n(\alpha,a,b)) - \sigma^{(\nu)}(\alpha))d\alpha. \tag{6.78}$$

Since $\zeta_0(x,a,b)$ satisfies

$$\zeta_0(x,a,b) = b - \int_x^a (\lambda_+^{(0)}(\alpha, \zeta_0(\alpha,a,b)) - \sigma^{(0)}(\alpha))d\alpha, \tag{6.79}$$

then

$$\begin{aligned}
&\zeta_{n+1}(x,a,b) - \zeta_0(x,a,b) \\
&\quad = \int_x^a (\lambda_+^{(\nu)}(\alpha, \zeta_n(\alpha,a,b)) - \lambda_+^{(0)}(\alpha, \zeta_0(\alpha,a,b)))d\alpha - \int_x^a (\sigma^{(\nu)}(\alpha) \\
&\quad\quad -\sigma^{(0)}(\alpha))d\alpha.
\end{aligned}$$

Now we prove that the limit of $\zeta_i(x,a,b)$ is $\eta^{(\nu)}(x,a,b)$, which satisfies Eq. (6.75). To this end we want to indicate that the inequality

$$|\zeta_i(x,a,b) - \zeta_0(x,a,b)| \le Ca(x-a) \tag{6.80}$$

holds for all i. Indeed, if (6.80) holds for $i = n$, then

$$\begin{aligned}
&\zeta_{n+1}(x,a,b) - \zeta_0(x,a,b) \\
&= \int_x^a (\lambda_+^{(0)}(\alpha,\zeta_n) - \lambda_+^{(0)}(\alpha,\zeta_0))d\alpha + \int_x^a (\lambda_+^{(\nu)}(\alpha,\zeta_n) - \lambda_+^{(0)}(\alpha,\zeta_n))d\alpha \\
&\qquad + \int_x^a (\sigma^{(\nu)}(\alpha) - \sigma^{(0)}(\alpha))d\alpha. \qquad (6.81)
\end{aligned}$$

According to the assumptions on $F^{(\nu)}$ we have

$$|\sigma^{(\nu)}(\alpha) - \sigma^{(0)}(\alpha)| \le C\alpha, \qquad (6.82)$$

$$\begin{aligned}
&|\lambda_+^{(\nu)}(\alpha,\zeta_n) - \lambda_+^{(0)}(\alpha,\zeta_n)| \\
&\le C\|\nabla\lambda\|_{L^\infty}(|r^{(\nu)}(\alpha,\zeta_n)| + |s^{(\nu)}(\alpha,\zeta_n) - s^{(0)}(\alpha,\zeta_n)|) \le C\alpha \qquad (6.83)
\end{aligned}$$

On the other hand,

$$\begin{aligned}
&\left|\int_x^a (\lambda_+^{(0)}(\alpha,\zeta_n) - \lambda_+^{(0)}(\alpha,\zeta_0))d\alpha\right| \\
&\le \left|\int_x^a (\lambda_{+s}^{(0)} \cdot s_y^{(0)})(\alpha,(\zeta_0 + \theta(\zeta_n - \zeta_0)) \cdot (\zeta_n - \zeta_0)d\alpha\right|.
\end{aligned}$$

By using the assumption Eq. (6.80) with $i = n$, we know $\zeta_0 + \theta(\zeta_n - \zeta_0)$ satisfies Eq. (6.61), then the remark of Lemma 6.7 implies

$$\left|\int_x^a (\lambda_+^{(0)}(\alpha,\zeta_n) - \lambda_+^{(0)}(\alpha,\zeta_0))d\alpha\right| \le \left(\log\frac{3}{2} + C\sqrt{a}\right) a(a-x). \qquad (6.84)$$

By substituting (6.82)–(6.84) into (6.81), we have Eq. (6.80) with $i = n+1$. Hence (6.80) holds for any i.

Now we prove the convergence of $\zeta_n(\alpha,a,b)$. In fact,

$$\begin{aligned}
&|\zeta_{n+1}(x,a,b) - \zeta_n(x,a,b)| \\
&= \left|\int_x^a (\lambda_+^{(\nu)}(\alpha,\zeta_n) - \lambda_+^{(\nu)}(\alpha,\zeta_{n-1}))d\alpha\right| \\
&\le \int_0^a |\partial_y\lambda_+^{(\nu)}(\alpha,(\zeta_{n-1} + \theta(\zeta_n - \zeta_{n-1})))|d\alpha \cdot \|\zeta_n - \zeta_{n-1}\|_{L^\infty}. \qquad (6.85)
\end{aligned}$$

The integral in the left side does not exceed

$$\begin{aligned}
&\int_0^a |\partial_y\lambda_+^{(0)}(\alpha,(\zeta_{n-1} + \theta(\zeta_n - \zeta_{n-1})))|d\alpha \\
&\quad + \int_0^a |(\partial_y\lambda_+^{(\nu)} - \partial_y\lambda_+^{(0)})(\alpha,(\zeta_{n-1} + \theta(\zeta_n - \zeta_{n-1})))|d\alpha. \qquad (6.86)
\end{aligned}$$

It is dominated by $\log\frac{3}{2}+C\sqrt{a}$ according to the remark of Lemma 6.7. Moreover, the assumption $F_{1,2}^{(\nu)}$ implies

$$|\partial_y\lambda_+^{(\nu)}-\partial_y\lambda_+^{(0)}|\le|\lambda_{+r}^{(\nu)}r_y^{(\nu)}+\lambda_{+s}^{(\nu)}s_y^{(\nu)}-\lambda_{+s}^{(0)}s_y^{(0)}|\le C(\alpha^{1/2}+\alpha^{-1/2}),$$

Then for small a

$$\int_0^a|\partial_y\lambda_+^{(\nu)}(\alpha,(\zeta_{n-1}+\theta(\zeta_n-\zeta_{n-1})))|d\alpha\le\log\frac{3}{2}+C\sqrt{a}\le\frac{1}{2}.$$

Substituting them into Eq. (6.85) we obtain the contraction of the sequence $\{\zeta_n\}$:

$$\|\zeta_{n+1}-\zeta_n\|_{L^\infty}\le\frac{1}{2}\|\zeta_n-\zeta_{n-1}\|_{L^\infty},\tag{6.87}$$

hence $\{\zeta_n\}$ is convergent. Obviously, the limit of $\{\zeta_n\}$ is the unique solution of Eq. (6.76). Meanwhile, the estimate Eq. (6.75) is obtained by taking limit in Eq. (6.80). □

Remark 6.1. Combining Lemma 6.11 with Eq. (6.56) we have

$$\eta_\pm^{(\nu)}(x,a,b)-b=\pm(\sqrt{a}(a-x)+O_1(|b|^{1/3}(a-x)))+O(a(a-x)),\tag{6.88}$$

which means that for $a>0$, the $\lambda_\pm$ leftward characteristics issuing from (a,b) can only leave $\Omega_\pm$ at $x=0$. The fact will be used in estimating the derivatives of $s^{(\nu+1)}$.

Lemma 6.13. *Under the above assumptions* $F_{1,2}^{(\nu)}$

$$|(s^{(\nu+1)}-s^{(0)})(x,y)|\le Cx,\tag{6.89}$$

$$|r^{(\nu+1)}(x,y)|\le Cx^{3/2}.\tag{6.90}$$

Proof. Denote $v(x,y)=(s^{(\nu+1)}-s^{(0)})(x,y)$, then v satisfies

$$\begin{cases}\partial_xv+(\lambda_+^{(\nu)}-\sigma^{(\nu)})\partial_yv=(-\lambda_+^{(\nu)}+\lambda_+^{(0)}+\sigma^{(\nu)}-\sigma^{(0)})\partial_ys^{(0)},\\ v(0,y)=0.\end{cases}\tag{6.91}$$

Integrating along the characteristics gives

$$|v(x,y)|\le\int_0^x|(-\lambda_+^{(\nu)}+\lambda_+^{(0)}+\sigma^{(\nu)}-\sigma^{(0)})(\partial_ys^{(0)})(\alpha,\eta(\alpha,a,b))|d\alpha.$$

By using the assumptions $F_{1,2}^{(\nu)}$ we know

$$|\lambda_+^{(\nu)}-\lambda_+^{(0)}|\le C\|\nabla\lambda\|_{L^\infty}(\|r^{(\nu)}\|_{L^\infty}+\|s^{(\nu)}-s^{(0)}\|_{L^\infty})\le C\alpha.$$

Besides, Theorem 6.2 indicates

$$|\partial_y s^{(0)}(\alpha, \eta(\alpha, x, y))| \le C(\alpha^3 + \eta^2)^{-1/3} \le C\alpha^{-1}.$$

Combining with Lemma 6.10 we obtain

$$|v(x, y)| \le C \int_0^x \alpha \cdot \alpha^{-1} d\alpha \le Cx.$$

Consider the estimate for $r^{(\nu+1)}$. It vanishes in $\Omega_0 \cup \Omega_+$. In Ω_-, $r^{(\nu+1)}$ is determined by an initial boundary value problem of the equation

$$\partial_x r^{(\nu+1)} + (\lambda_-^{(\nu)} - \sigma^{(\nu)})\partial_y r^{(\nu+1)} = 0. \tag{6.92}$$

The initial condition on $x = 0$ is $r^{(\nu+1)} = 0$, while the boundary condition on $y = 0$ is derived from the Rankine-Hugoniot conditions. Denote by $\ell_- : \ y = \eta_-(x, a, b)$ the λ_- characteristics passing through (a, b). If the curve intersects with $x = 0$ at $(\xi, 0)$, then Lemma 6.9 implies

$$|r^{(\nu+1)}| \le C|[s^{(\nu+1)}]^3|. \tag{6.93}$$

On the other hand, Lemma 6.6 indicates $[s^{(0)}] \le Cx^{1/2}$. Hence combining with the estimate (6.89) we have

$$[s^{(\nu+1)}] \le Cx^{1/2}.$$

Substituting it into Eq. (6.93) implies the validity of Eq. (6.90) on $y = 0$. Then integrating Eq. (6.92) along the λ_- characteristics shows the validity of Eq. (6.90) in the whole Ω_-. □

Lemma 6.14. *Under the above assumptions on* $F_{1,2}^{(\nu)}$

$$|\partial_{x,y}(s^{(\nu+1)} - s^{(0)})(x, y)| \le C(x^3 + y^2)^{-1/6}, \tag{6.94}$$

$$|\partial_{x,y}(r^{(\nu+1)})(x, y)| \le C\sqrt{x}. \tag{6.95}$$

Proof. Let $v = \partial_y(s^{(\nu+1)} - s^{(0)})(x, y)$, then it satisfies

$$\begin{aligned}\partial_x v + (\lambda_+^{(\nu)} - \sigma^{(\nu)})\partial_y v = (&-\lambda_+^{(\nu)} + \lambda_+^{(0)} + \sigma^{(\nu)} - \sigma^{(0)})s_{yy}^{(0)} \\ &-\partial_y \lambda_+^{(\nu)} v + \partial_y(\lambda_+^{(\nu)} - \lambda_+^{(0)})s_y^{(0)}.\end{aligned} \tag{6.96}$$

By using Theorem 6.2 we have

$$\begin{aligned}
&\left|\int_0^a (-\lambda_+^{(\nu)} + \lambda_+^{(0)} + \sigma^{(\nu)} - \sigma^{(0)}) s_{yy}^{(0)} d\alpha\right| \\
&\quad \le \int_0^a C\alpha(\alpha^3 + \eta^2)^{-5/6} d\alpha \\
&\quad \le Ca^2(a^3 + b^2)^{-5/6} \le C(a^3 + b^2)^{-1/6}. \\
&\left|\int \partial_y(\lambda_+^{(\nu)} - \lambda_+^{(0)}) s_y^{(0)} d\alpha\right| \\
&\quad \le \int |(\lambda_{+r}^{(\nu)} r_y^{(\nu)} + \lambda_{+s}^{(\nu)} s_y^{(\nu)} - \lambda_{+s}^{(0)} s_y^{(0)})| \cdot |s_y^{(0)}| d\alpha \\
&\quad \le \int |(\lambda_{+r}^{(\nu)} r_y^{(\nu)} + \lambda_{+s}^{(\nu)} (s_y^{(\nu)} - s_y^{(0)} + (\lambda_{+s}^{(\nu)} - \lambda_{+s}^{(0)}) s_y^{(0)}| \cdot |s_y^{(0)}| d\alpha \\
&\quad \le C\int_0^a |(\alpha^3 + \eta^2)^{-1/6} + \alpha^{-1/2} + \alpha(\alpha^3 + \eta^2)^{-1/3}| \cdot (\alpha^3 + \eta^2)^{-1/3} d\alpha \\
&\quad \le C\int_0^a \alpha^{-1/2} d\alpha \cdot (a^3 + b^2)^{-1/3} \le C(a^3 + b^2)^{-1/6}.
\end{aligned}$$

Hence we established

$$|v| \le \int_0^a g(\alpha)|v| d\alpha + h(a), \tag{6.97}$$

where $h(a) \le C(a^3 + b^2)^{-1/6}$. Besides, Lemma 6.7 implies

$$\int_0^a g(\alpha) d\alpha \le \log\frac{3}{2} + C\sqrt{a},$$

then for small a the estimate of the derivatives with respect to y in Eq. (6.94) can be obtained. Similarly, we can establish the estimate of the derivatives with respect to x.

Consider the estimate of derivatives of $r^{(\nu+1)}$. From Eq. (6.66) we have

$$r^{(\nu+1)}(x) = f(s_+^{(\nu+1)}(x), s_-^{(\nu+1)}(x))[s^{(\nu+1)}]^3.$$

Taking derivatives with respect to x and applying the estimate (6.94) we obtain

$$\begin{aligned}
|r_x^{(\nu+1)}| &\le C(3[s^{(\nu+1)}]^2|[s_x^{(\nu+1)}]| + [s^{(\nu+1)}]^3(s_{-x}^{(\nu+1)} + |s_{+x}^{(\nu+1)}|)) \\
&\le C\sqrt{x}.
\end{aligned} \tag{6.98}$$

As did before, by using the integration along λ_- characteristics we can establish the estimate of $r_x^{(\nu+1)}$ in the whole domain Ω.

Finally, in view of that $|\lambda_-^{(\nu)} - \sigma^{(\nu)}|$ has positive lower bound, the estimate of derivative of $r^{(\nu+1)}$ with respect to y can also be obtained. □

By using the conclusion of Lemma 6.11 to Lemma 6.14, we obtained the validity of $F_{1,2}^{(\nu)}$ for any ν by induction.

Proof of Theorem 6.3. The key point of the proof of Theorem 6.3 is to establish the convergence of the sequence of $\{r^{(\nu)}, s^{(\nu)}\}$. Next we are going to give the estimates of $|\sigma^{(\nu)} - \sigma^{(\nu-1)}|, |s^{(\nu+1)} - s^{(\nu)}|, |r^{(\nu+1)} - r^{(\nu)}|$ one by one.

$$\begin{aligned}|\sigma^{(\nu)} - \sigma^{(\nu-1)}| \le &\frac{1}{2}\|\lambda'_+\|_{L^\infty}\|s^{(\nu)} - s^{(\nu-1)}\|_{L^\infty}(1 + O(x)) \\ &+C\|r^{(\nu)} - r^{(\nu-1)}\|_{L^\infty},\end{aligned} \tag{6.99}$$

$$\|s^{(\nu+1)} - s^{(\nu)}\|_{L^\infty} \le \frac{9}{10}(\|r^{(\nu)} - r^{(\nu-1)}\|_{L^\infty} + \|s^{(\nu)} - s^{(\nu-1)}\|_{L^\infty}), \tag{6.100}$$

$$\|r^{(\nu+1)} - r^{(\nu)}\|_{L^\infty} \le \frac{1}{20}(\|r^{(\nu)} - r^{(\nu-1)}\|_{L^\infty} + \|s^{(\nu)} - s^{(\nu-1)}\|_{L^\infty}), \tag{6.101}$$

First, like Lemma 6.11 we have

$$|\sigma^{(\nu)} - \sigma^{(\nu-1)}| \le C(\|s^{(\nu)} - s^{(\nu-1)}\|_{L^\infty} + \|r^{(\nu)} - r^{(\nu-1)}\|_{L^\infty}). \tag{6.102}$$

To establish more precise estimate we write

$$\begin{aligned}&|\sigma^{(\nu)} - \sigma^{(\nu-1)}| = |H(r^{(\nu)}, s^{(\nu)} - H(r^{(\nu-1)}, s^{(\nu-1)}| \\ &\le C|r^{(\nu)} - r^{(\nu-1)}| + |H(0, s^{(\nu)}) - H(0, s^{(\nu-1)})| \\ &+|(H(r^{(\nu)}, s^{(\nu)}) - H(r^{(\nu)}, s^{(\nu-1)})) - (H(0, s^{(\nu)}) - H(0, s^{(\nu-1)}))|.\end{aligned} \tag{6.103}$$

By using the notations of Eq. (6.52)

$$\begin{aligned}H(0, s^{(\nu)}) - H(0, s^{(\nu-1)}) &= \frac{[f(m)(s^{(\nu)})]}{[m(s^{(\nu)})]} - \frac{[f(m)(s^{(\nu-1)})]}{[m(s^{(\nu-1)})]} \\ &= \left(\frac{d}{dm}\frac{[f(m)]}{[m]}\right)_{m=m^*}\left(\frac{dm}{ds}\right)_{s=s^*}(s^{(\nu)} - s^{(\nu-1)}) \\ &= \left(\frac{[m]f'(m) - [f(m)]}{[m]^2}\right)_{m=m^*}\left(\frac{dm}{ds}\right)_{s=s^*}(s^{(\nu)} - s^{(\nu-1)}),\end{aligned}$$

where $s^* = s^{(\nu-1)} + \theta(s^{(\nu)} - s^{(\nu-1)})$, $0 \le \theta \le 1, m^* = m(s^*)$. Then

$$\begin{aligned}&|H(0, s^{(\nu)}) - H(0, s^{(\nu-1)})| \\ &\le |\frac{1}{2}f''(m_0) + O([m])| \cdot \left|\frac{dm}{ds}\right| \cdot |s^{(\nu)} - s^{(\nu-1)}| \\ &= \left(\frac{1}{2}\lambda'_+(s_0) + O(|s^{(\nu)} - s^{(0)}|)\right)|s^{(\nu)} - s^{(\nu-1)}|,\end{aligned}$$

Substituting it into Eq. (6.103) implies Eq. (6.99).

To prove Eq. (6.100), we denote $v = s^{(\nu+1)} - s^{(\nu)}$, then v satisfies

$$\partial_x v + (\lambda_+^{(\nu)} - \sigma^{(\nu)})\partial_y v = (\lambda_+^{(\nu-1)} - \lambda_+^{(\nu)} + \sigma^{(\nu)} - \sigma^{(\nu-1)})\partial_y s^{(\nu)}.$$

Integrating along characteristics gives

$$|v| \le \int_0^a |(\lambda_+^{(\nu-1)} - \lambda_+^{(\nu)} + \sigma^{(\nu)} - \sigma^{(\nu-1)})\partial_y (s^{(\nu)} - s^{(0)})(\alpha, \eta(\alpha, a, b))|d\alpha$$
$$+ \int_0^a |(\lambda_+^{(\nu-1)} - \lambda_+^{(\nu)}) + (\sigma^{(\nu)} - \sigma^{(\nu-1)})\partial_y s^{(0)}(\alpha, \eta(\alpha, a, b)))|d\alpha. \quad (6.104)$$

Combining with Lemma 6.8 we obtain

$$|v| \le \left(\frac{3}{2}\log\frac{3}{2} + C\sqrt{a}\right)(\|r^{(\nu)} - r^{(\nu-1)}\|_{L^\infty} + \|s^{(\nu)} - s^{(\nu-1)}\|_{L^\infty})$$

Due to $\frac{3}{2}\log\frac{3}{2} < \frac{3}{4}$, then Eq. (6.100) holds, as a is small.

Finally, let $v = r^{(\nu+1)} - r^{(\nu)}$, then v satisfies

$$\partial_x v + (\lambda_-^{(\nu)} - \sigma^{(\nu)})\partial_y v = (\lambda_-^{(\nu-1)} - \lambda_-^{(\nu)} + \sigma^{(\nu)} - \sigma^{(\nu-1)})\partial_y r^{(\nu)}, \quad (6.105)$$

and $v = 0$ on $x = 0$. Integrate it along λ_- characteristics. If the leftward characteristics first meet x-axis before it meet y-axis, then

$$|v| \le |r^{(\nu+1)}(x, 0_-) - r^{(\nu)}(x, 0_-)|$$
$$+ \int_x^a |(\lambda_-^{(\nu-1)} - \lambda_-^{(\nu)})\partial_y r^{(\nu)}(\alpha, \eta(\alpha, a, b))|d\alpha$$
$$+ \int_x^a |(\sigma^{(\nu-1)} - \sigma^{(\nu)})\partial_y r^{(\nu)}(\alpha, \eta(\alpha, a, b))|d\alpha. \quad (6.106)$$

Otherwise, the first term in the right hand side of Eq. (6.106) does not appear, and the lower bound of the integral should be 0. Due to $|\partial_y r^{(\nu)}| \le C\sqrt{a}$, the integral in Eq. (6.106) is dominated by

$$Ca^{3/2}(\|r^{(\nu)} - r^{(\nu-1)}\|_{L^\infty} + \|s^{(\nu)} - s^{(\nu-1)}\|_{L^\infty}).$$

Moreover, Eq. (6.66) indicates

$$|r^{(\nu+1)}(x, 0_-) - r^{(\nu)}(x, 0_-)| \le C|[s^{(\nu+1)}]^3 - [s^{(\nu)}]^3| \le Ca\|s^{(\nu+1)} - s^{(\nu)}\|_{L^\infty}.$$

Combining all these facts we obtain

$$|(r^{(\nu+1)} - r^{(\nu)})(a, b)| \le Ca^{3/2}(\|r^{(\nu)} - r^{(\nu-1)}\|_{L^\infty} + \|s^{(\nu)} - s^{(\nu-1)}\|_{L^\infty})$$
$$+ Ca\|s^{(\nu+1)} - s^{(\nu)}\|_{L^\infty}.$$

Therefore, (ref700) holds, as a is small.

By adding the estimates shown from Eq. (6.99) to Eq. (6.101) we obtain

$$\|r^{(\nu+1)} - r^{(\nu)}\|_{L^\infty} + \|s^{(\nu+1)} - s^{(\nu)}\|_{L^\infty})$$
$$\le \frac{19}{20}(\|r^{(\nu)} - r^{(\nu-1)}\|_{L^\infty} + \|s^{(\nu)} - s^{(\nu-1)}\|_{L^\infty}). \quad (6.107)$$

Then we obtain the convergence of the sequence $\{r^{(\nu)}(x,y), s^{(\nu)}(x,y)\}$. Correspondingly, the convergence of $\{\sigma^{(\nu)}\}$ is also obtained. Denote the limit of these sequences by $r(x,y), s(x,y), \sigma(x)$ respectively, then $(r(x,y), s(x,y))$ is continuous, and satisfies the corresponding integral equation. Furthermore, it is easy to verify that $(r(x,y), s(x,y), \sigma(x))$ satisfies the differential system

$$\begin{cases} \partial_x r(x,y) + (\lambda_- - \sigma)\partial_y r(x,y) = 0, \\ \partial_x s(x,y) + (\lambda_+ - \sigma)\partial_y s(x,y) = 0, \end{cases} \tag{6.108}$$

as well as the Rankine-Hugoniot condition on $y = \phi(x)$, where $\phi(x) = \int_0^x \sigma(\alpha)d\alpha$. Returning to the original coordinates we obtain the weak solution of Eq. (6.42). By taking the limit for the estimates satisfied by $\{r^{(\nu)}(x,y), s^{(\nu)}(x,y)\}$ and $\sigma^{(\nu)}(x)$ we obtain the estimates in Theorem 6.3.

The main theorem in this section is thus proved. □

6.2.4 *The case for full Euler system*

If people wants to have a ore precise description on shock formation in gas dynamics, they should study the full Euler system. For the two-dimensional steady flow the system governing the flow is

$$\frac{\partial}{\partial x}\begin{pmatrix} \rho u \\ p+\rho u^2 \\ \rho uv \\ \rho uE + pu \end{pmatrix} + \frac{\partial}{\partial y}\begin{pmatrix} \rho v \\ \rho uv \\ p+\rho v^2 \\ \rho vE + pv \end{pmatrix} = 0, \tag{6.109}$$

where (u,v), p, ρ represent components of velocity, pressure, density as before, and E is energy respectively. For polytropic gas $E = q^2/2 + e, p = A(S)\rho^\gamma$, where S is entropy, γ is adiabatic exponent, $q^2 = u^2+v^2$, $e = p/((\gamma-1)\rho)$, $A(S) = (\gamma-1)e^{c_v(S-S_0)}$ with c_v being a constant.

From (6.109) we can derive a Bernoulli relation satisfied by flow parameters

$$\frac{1}{2}(u^2+v^2) + \frac{\gamma p}{(\gamma-1)\rho} = const. \tag{6.110}$$

Let θ be the angle of velocity, $\theta = \arctan v/u$, then Eq. (6.109) can be written as

$$\begin{cases} K(p,\theta,S)\dfrac{dp}{d\ell_+} + \dfrac{d\theta}{d\ell_+} = 0, \\ \dfrac{dS}{d\ell_0} = 0, \\ K(p,\theta,S)\dfrac{dp}{d\ell_-} - \dfrac{d\theta}{d\ell_-} = 0, \end{cases} \tag{6.111}$$

where $\dfrac{d}{d\ell_{\pm,0}} = \dfrac{\partial}{\partial x} + \lambda_{\pm,0}\dfrac{\partial}{\partial y}$ is directional derivative along the characteristics $\ell_{\pm,0}$, $K(p,\theta,S) = \dfrac{1}{\rho q^2 \tan A}$, A is the Mach angle,

$$\lambda_\pm = \frac{uv \pm \sqrt{q^2-a^2}}{u^2-a^2}, \quad \lambda_0 = \frac{v}{u}.$$

Like the discussion for the irrotational case we also introduce new unknown functions $z = \theta - F(q), w = \theta + F(q)$ with $F(q) = \displaystyle\int \frac{\sqrt{q^2-a^2}}{aq}dq$, which amount to Riemannian invariants in irrotational case. Then takes the form

$$\begin{cases} \dfrac{dz}{d\ell_+} - d\dfrac{dS}{d\ell_+} = 0, \\ \dfrac{dS}{d\ell_0} = 0, \\ \dfrac{dw}{d\ell_-} + d\dfrac{dS}{d\ell_-} = 0, \end{cases} \tag{6.112}$$

where $d = \dfrac{1}{\gamma(\gamma-1)A(s)}\dfrac{a\sqrt{q^2-a^2}}{q^2}$. According to the shock theory of compressible flow ([46]), across λ_+ shock the jump of z is the main part of the jump of parameters of the flow, while the jump of w and S is only a quantity of third order, comparing the strength of shock.

For the formation of shock we have:

Theorem 6.4. *Assume that a uniform upstream flow with velocity $(u_0, 0)$ moves above the wall $y = f(x)$ from left to right, $u_0 > a_0$, $f(x)$ satisfies*

$$\begin{cases} f(x) \equiv 0, & x \le 0, \\ f(x) > 0, f'(x) > 0, f''(x) > 0, & x > 0, \end{cases}$$

and the envelop of the family of characteristics ℓ_+ has a cusp at (x_0, y_0), then the system Eq. (6.109) admits a solution, which is continuously differentiable in $x < x_0$ and has a shock $\Gamma : y = \phi(x)$ starting from the point (x_0, y_0). In the neighborhood Ω of (x_0, y_0), the solution is also continuously differentiable in $\Omega \setminus \Gamma$. Besides, the solution satisfies Rankine-Hugoniot condition and entropy conditions on Γ, and the following estimates in Ω hold.

$$\begin{cases} \phi(x) = y_0 + \alpha(x-x_0) + O((x-x_0)^2), \\ w = w_0 + O((x-x_0)^{3/2}), \\ z = z_0 + O((x-x_0)^3 + (y-y_0-\alpha(x-x_0))^2)^{1/6}, \\ S = S_0 + O((x-x_0)^{3/2}), \end{cases} \tag{6.113}$$

where w_0, z_0, S_0 *are the value of the corresponding functions at the point* (x_0, y_0), α *is the slope of the* λ_+ *characteristics at this point.*

The proof of the theorem can be found in [46]. As shown there, since the system has three families of characteristics, then together with the formation of shock, other weaker singularities of the solutions of Eq. (6.109) may also propagate into the domain $x > x_0$.

Appendix A

Brief review on paradifferential operators

In the discussion of various problems on nonlinear partial differential equations, a good method of linearization often plays crucial role. The situation also likes this in the analysis of singularities for nonlinear equations. Based on the diadic decompositions J.M.Bony established the theory of paralinearization and paradifferential operators, which is an powerful tool in the study of propagation of singularities. The main reason is that in the process of such linearizations people can pay more attentions on the singular terms, while the remainders with higher regularity are often given up. Since such a "give up" does not influence the analysis on the part containing the main singularities of solutions, then a nonlinear problem can be reduced to a linear problem, which is easier to be treated. Later, S.Alinhac also developed the theory of paracomposition, which can also be applied to describe the propagation of conormal distributions. To the readers convenience we shall give a brief review on the main results of the theory of paradifferential operators without proof and more explanation. Readers can find those in the references [3], [19].

A.1 Diadic decomposition

Denote by $\mathscr{B}(0,1) \subset R^n$ the unit ball with center at the origin. By taking $\kappa > 1$, we make a set of diadic rings:

$$\mathscr{C}_j = \{\xi \in R^n;\ \kappa^{-1}2^j \le |\xi| \le \kappa 2^{j+1}\}, \tag{A.1}$$

where j is an integer, $\xi = (\xi_1, \cdots, \xi_n), |\xi| = (\sum \xi_j^2)^{1/2}$, then $\mathscr{B}(0,1) \bigcup \left(\bigcup_{j=0}^{\infty} \mathscr{C}_j\right)$ covers the whole R^n.

Theorem A.1. *There are functions* $\psi(\xi), \phi(\xi) \in C_0^\infty$, $0 \le \psi, \phi \le 1$, *such* *that*

(1) $\text{supp}\psi \subset \mathscr{B}(0,1)$, $\text{supp}\phi \subset \mathscr{C}_0$,

(2) $\psi(\xi) + \sum_{j=0}^{\infty} \phi(2^{-j}\xi) = 1$, $\forall \xi \in R^n$,

(3) for any integer ℓ,

$$\psi(\xi) + \sum_{j=0}^{\ell-1} \phi(2^{-j}\xi) = \psi(2^{-\ell}\xi), \ \forall \xi \in R^n.$$

Definition A.1. For any distribution $u \in \mathscr{S}'(R^n)$, one can define its diadic decomposition (or Littlewood-Paley decomposition):

$$u = \sum_{j=-1}^{\infty} u_j(x), \tag{A.2}$$

where $u_j(x)$ is determined by

$$\hat{u}_{-1} = \psi(\xi)\hat{u}(\xi), \quad \hat{u}_j(\xi) = \phi(2^{-j}\xi)\hat{u}(\xi) \quad (j \geq 0).$$

ψ, ϕ are the functions given in Theorem A.1.

Theorem A.2. *If* $u \in H^s(R^n)$ *has a diadic decomposition, then there are constants* c_j *and* C*, such that*

$$\|u_j\|_0 \leq c_j 2^{-j\nu}, \quad j = -1, 0, 1, \cdots \tag{A.3}$$

with

$$\left(\sum_{j=-1}^{\infty} c_j^2\right)^{1/2} \leq C\|u\|_s. \tag{A.4}$$

Theorem A.3. *For any real s, the following propositions are equivalent.*

(1) $u \in H^s(R^n)$,

(2) u can be decomposed as $\sum_{j=-1}^{\infty} u_j(x)$*, where* u_j *satisfies*

$$\text{supp}\hat{u_j} \subset \mathscr{B}(0, \kappa 2^j), \quad \kappa > 1,$$

$$\|u_j\|_0 \leq 2^{-\nu} c_j, \quad \sum_{j=-1}^{\infty} c_j^2 < \infty. \tag{A.5}$$

(3) There is a positive integer $m > s$*, such that u can be written as* $\sum_{j=-1}^{\infty} u_j(x)$ *with* $u_j \in C^\infty$*, and for any* $|\lambda| \leq m$,

$$\|D^\lambda u_j\|_0 \leq c_{j\lambda} \cdot 2^{-js+j|\lambda|}, \quad \sum_{j=-1}^{\infty} c_{j\lambda}^2 < \infty. \tag{A.6}$$

Theorem A.4. *If $\rho > 0$ is not an integer, $u \in C^\rho(R^n)$ has a diadic decomposition, then there is constant C, such that*

$$\|u_j\|_{L^\infty} \le C2^{-j\nu}\|u\|_{C^\rho}, \quad j = -1, 0, 1, \cdots \tag{A.7}$$

Theorem A.5. *For any non-integer $\rho > 0$, the following propositions are equivalent.*

(1) $u \in C^\rho(R^n)$,

(2) u can be decomposed as $\sum_{j=-1}^{\infty} u_j(x)$, where u_j satisfies

$$\text{supp}\hat{u}_j \subset \mathscr{C}_j, \quad \|u_j\|_{L^\infty} \le C \cdot 2^{-j\rho},$$

(3) u can be decomposed as $\sum_{j=-1}^{\infty} u_j(x)$, where u_j satisfies

$$\text{supp}\hat{u}_j \subset \mathscr{B}(0, \kappa 2^j)\ (\kappa > 1), \quad \|u_j\|_{L^\infty} \le C \cdot 2^{-j\rho},$$

(4) there is a positive integer $m > \rho$, such that u can be written as $\sum_{j=-1}^{\infty} u_j(x)$ with $u_j \in C^\infty$, and for any $|\lambda| \le m$,

$$\|D^\lambda u_j\|_{L^\infty} \le C_\lambda 2^{-j\rho + j|\lambda|}. \tag{A.8}$$

When the index ρ in Theorem A.5 takes integer, the Hölder condition should be replaced by the Zygmund condition, which is $u \in L^\infty$ and

$$|u(x+y) + u(x-y) - 2u(x)| \le M|y|, \quad |y| > 0. \tag{A.9}$$

Theorem A.6. *The following propositions are equivalent.*

(1) $u \in L^\infty(R^n)$ and

$$\sup_{|y|>0} |y|^{-1} \cdot |u(x+y) + u(x-y) - 2u(x)| < \infty,$$

(2) u can be decomposed as $\sum_{j=-1}^{\infty} u_j(x)$, where u_j satisfies

$$\text{supp}\hat{u}_j \subset \mathscr{C}_j, \quad \|u_j\|_{L^\infty} \le C \cdot 2^{-j},$$

(3) u can be decomposed as $\sum_{j=-1}^{\infty} u_j(x)$, where u_j satisfies

$$\text{supp}\hat{u}_j \subset \mathscr{B}(0, \kappa 2^j)\ (\kappa > 1), \quad \|u_j\|_{L^\infty} \le C \cdot 2^{-j},$$

(4) there is a positive integer $m > \rho$, such that u can be written as $\sum_{j=-1}^{\infty} u_j(x)$ with $u_j \in C^\infty$, and for any $|\lambda| \le m$,

$$\|D^\lambda u_j\|_{L^\infty} \le C_\lambda 2^{-j + j|\lambda|}. \tag{A.10}$$

By using the diadic decomposition one can also study the local or microlocal property of functions.

Theorem A.7. *Assume that s, s' are real numbers, $s' \geq s$, then the following statements are equivalent:*

(1) $u \in H^s_{x_0} \cap H^s_{(x_0,\xi_0)}$,

(2) there exists $\phi(x) \in C_0^\infty(R^n)$, which is 1 in a neighborhood of x_0, and for the given ϕ there is a conical neighborhood Γ, such that

$$\phi u = \sum_{j=-1}^{\infty} v_j + \sum_{j=-1}^{\infty} v'_j,$$

where

$$\|v_j\|_0 \leq c_j 2^{-js}, \quad \mathrm{supp}\hat{v}_j \subset \mathscr{C}_j,$$

$$\|v'_j\|_0 \leq c_j 2^{-js'}, \quad \mathrm{supp}\hat{v}'_j \subset \mathscr{C}_j \subset \Gamma^c, \quad \sum c_j^2 < \infty.$$

Denote by $S_k u$ the partial sum $\sum_{j=-1}^{k-1} u_j$ of the diadic decomposition of the function u, i.e.

$$S_k u = \psi(D)u + \sum_{j=-1}^{k-1} \phi(2^{-j}D)u,$$

then $S_k u \in \mathscr{S}(R^n)$. Meanwhile, if $u \in H^s$ or $u \in C^\rho$, $S_k u$ has limit u in H^s or C^ρ. Therefore, $S_k u$ is a smooth approximation of a non-smooth function u. Moreover,

Theorem A.8. *If $u \in H^s$ with $s > 0$, then*

$$\|S_k u - u\|_0 \leq C \cdot 2^{-ks}\|u\|_s. \tag{A.11}$$

If $u \in C^\rho$ with $\rho > 0$, then

$$\|S_k u - u\|_{L^\infty} \leq C \cdot 2^{-k\rho}\|u\|_{C^\rho}. \tag{A.12}$$

Theorem A.9. *If $u \in H^s$,* supp$u \subset F$, *then for any $N > 0$*

$$\|2^{js}(1 + 2^j d(x,F))^N u_j(x)\| \leq c_{jN}, \quad \sum_j c_{jN}^2 < \infty, \tag{A.13}$$

where $d(x,F)$ is the distance from x to F.

A.2 Paradifferential operators and paralinearzation

Next we give the definition of paradifferential operators and paraproduct. Paradifferential operators can be defined by using diadic composition, or by a special integral of Fourier transformations. They can also be defined as pseudodifferential operators of $OpS^0_{1,1}$ class. In the sequel we use the first method.

Definition A.2. Let the diadic decompositions of $a(x), u(x)$ be $a = \sum a_j$, $u = \sum u_j$ respectively. For any integer N, define

$$T_a u = \sum_{q=N-1}^{\infty} \sum_{p=-1}^{q-N} a_p u_q, \tag{A.14}$$

then $T_a u$ is called the **paraproduct** of u with a.

Theorem A.10. *Assume that σ, s are real numbers, $a \in L^\infty$, then T_a is an bounded operator $C^\sigma \to C^\sigma$ or $H^s \to H^s$. Moreover, the norm of the operator satisfies $\|T_a\| \le C\|a\|_{L^\infty}$*

The number N in Definition A.2 and the constant κ, the functions ϕ, ψ introduced by the diadic decomposition may cause some difference of $T_a u$. When $a \in C^\rho$, the difference is a bounded operator $C^\sigma \to C^{\sigma+\rho}$ or $H^s \to H^{s+\rho}$. Such operators are called **ρ-regular operator**.

Theorem A.11. *Denote*

$$r(a,u) = au - T_a u - T_u a, \quad r'(a,u) = au - T_a u, \tag{A.15}$$

(1) If $a \in C^\rho(R^n)$, $\rho > 0, u \in C^\sigma(R^n)$, then

(a) If $\sigma \ge 0$, then $\|r(a,u)\|_{C^{\rho+\sigma}} \le C\|a\|_{C^\rho}\|u\|_{C^\sigma}$.

(b) If $-\rho < \sigma < 0$, then $\|r'(a,u)\|_{C^{\rho+\sigma}} \le C\|a\|_{C^\rho}\|u\|_{C^\sigma}$.

(c) If $\sigma = 0$, then $\|r'(a,u)\|_{C^{\rho+\sigma-\epsilon}} \le C\|a\|_{C^\rho}\|u\|_{C^\sigma}$ with ϵ being any small positive number.

(2) If $a \in H^t(R^n)$, $t > n/2, u \in H^s(R^n)$, then

(a) If $s \ge n/2$, then $\|r(a,u)\|_{t+s-n/2} \le C\|a\|_s\|u\|_s$.

(b) If $-t + n/2 < s < n/2$, then $\|r'(a,u)\|_{t+s-n/2} \le C\|a\|_t\|u\|_s$.

(c) If $s = n/2$, then $\|r'(a,u)\|_{t-\epsilon} \le C\|a\|_s\|u\|_s$ with $\epsilon > 0$ being any small positive number.

Theorem A.12. *Assume $a \in C^\rho(R^n), b \in C^\rho(R^n), \rho > 0$, then the operator $T_a T_b - T_{ab}$ is a ρ-regular operator, and $\|T_a T_b - T_{ab}\| \le C\|a\|_{C^\rho}\|b\|_{C^\rho}$.*

Theorem A.13. *Assume $a \in C^\rho(R^n), \rho > 0$, then the conjugate operator T_σ^* of the operator T_a is H^s bounded. Moreover, the operator $T_a - T_a^*$ is a bounded operator from H^s to $H^{s+\rho}$, satisfying $\|T_a - T_a^*\| \le C\|a\|_{C^\rho}$.*

Theorem A.14. *Assume $a, b \in C^{\rho,k}, \rho > 0$, k is a non-negative integer, then*

(1) T_a is a linear continuous map $C^{\sigma,k} \to C^{\sigma,k}$ and $H^{s,k} \to H^{s,k}$,

(2) $R = T_aT_b - T_{ab}$ is a linear continuous map $C^{\sigma,k} \to C^{\sigma+\rho,k}$ and $H^{s,k} \to H^{s+\rho,k}$.

Theorem A.15. *Assume $a \in C^\rho(R^n), \rho > 0$, $v(x') \in C^\sigma(R^{n-1})$ (or $H^s(R^{n-1})$), then*

$$(T_a v)_{x_n=0} = T_{(a|_{x_n=0})}v + Rv, \tag{A.16}$$

where v in the left-hand side is regular as a C^σ function independent of x_n, R is ρ-regular operator.

Theorem A.16. *If $a \in C^{|\alpha|+\rho}$ with $\rho > 0$, then the Leibniz formula for paraproduct*

$$D^\alpha(T_a u) = \sum_{\alpha'+\alpha''\le\alpha} \frac{\alpha!}{\alpha'!\alpha''!} T_{D^{\alpha'}a} D^{\alpha''} u \tag{A.17}$$

holds.

On the commutator of paraproduct with mollifier the following theorem holds.

Theorem A.17. *Assume $a \in C^\rho(R^n), \rho > 1$, $u \in H^s(R^n)$, then*

(1) $[T_a, J_\epsilon]u \in H^{s+1}(R^n)$, and

$$\|[T_a, J_\epsilon]u\|_{s+1} \le C\|u\|_s, \tag{A.18}$$

$$\|[T_a, J_\epsilon]u\|_s \le C\epsilon\|u\|_s, \tag{A.19}$$

where C is independent of ϵ, s.

(2) $[T_a, J_\epsilon]u \to 0$ in H^{s+1}, as $\epsilon \to 0$.

Assume that $\ell(x,\xi)$ is a homogeneous function of degree m of $\xi \in R^n$, being C^∞ for $\xi \ne 0$. As a function of x, the derivatives of ℓ with respect to x belongs to C^ρ $(\rho > 0)$, then $\ell(x,\xi)$ has a spherical harmonic decomposition

$$\ell(x,\xi) = \sum a_\nu(x)h_\nu(\xi), \tag{A.20}$$

where $a_\nu(x) \in C^\rho$, $h_\nu(\xi)$ is a homogeneous function of degree m, which is C^∞ as $\xi \neq 0$.

Definition A.3. For the function $\ell(x,\xi)$ given as above, we can define a linear operator by using its spherical harmonic decomposition (A.20)

$$T_\ell u = \sum_\nu T_{a_\nu} h_\nu(D) s(D) u, \tag{A.21}$$

where T_{a_ν} is the paraproduct operator introduced by a_ν, and $h_\nu(D), s(D)$ are pseudodifferential operators with symbols $h_\nu(\xi), s(\xi)$, the function $s(\xi) \in C^\infty$ is equal to 0 in a neighborhood of the origin, and is equal to 1 for large $|\xi|$. The operator defined by Eq. (A.21) is called **paradifferential operator** with symbol $\ell(x,\xi)$.

The function $\ell(x,\xi)$ in the above definition can be replaced by a sum of homogeneous functions with different degrees. If the highest degree in the sum is m in the sum, then one denotes $\ell \in \Sigma^m_\rho(\Omega)$. In accordance, the corresponding paradifferential operator L belongs to the class $Op\Sigma^m_\rho(\Omega)$. The term with highest degree in the sum $\ell(x,\xi)$ is called **principal symbol**.

Theorem A.18. *The paradifferential operator defined by Eq. (A.21) is a linear continuous map $H^s \to H^{s-m}$ or $C^\rho \to C^{\rho-m}$. When the function $s(\xi)$ in Eq. (A.21) or κ, $\psi(\xi)$,$\phi(\xi)$ are replaced by corresponding functions with same property, the error caused by such a replacing is a $(\rho-m)$-regular operator.*

Theorem A.19. *Assume that $h(\xi)$ is a C^∞ function, vanishes in a neighborhood of the origin. It is a homogeneous function of degree m for large $|\xi|$. Let $a(x) \in C^\rho$ with $\rho > m$, then*

$$h(D)T_a u = \sum_{|\alpha| \le [\rho]} \frac{1}{\alpha!} T_{D^\alpha a} h^{(\alpha)}(D) u + Ru, \tag{A.22}$$

where R is a $(\rho - m)$-regular operator.

Theorem A.20. *Assume that $\ell_1(x,\xi)$,$\ell_2(x,\xi)$ satisfy the condition in Definition A.3, they are homogeneous function of degree m_1, m_2 respectively. Let*

$$\ell(x,\xi) = (\ell_1 \# \ell_2)(x,\xi) = \sum_{|\alpha| \le [\rho]} \frac{1}{\alpha!} \partial^\alpha_\xi \ell_1 D^\alpha_x \ell_2, \tag{A.23}$$

then $T_{\ell_1} T_{\ell_2} = T_\ell + R$ with R being a $(\rho - m_1 - m_2)$-regular operator.

Theorem A.21. *Assume that $\ell(x,\xi)$ satisfies the conditions in Definition A.3, let*

$$\ell^*(x,\xi) = \sum_{|\alpha|\le[\rho]} \frac{1}{\alpha!}\partial_\xi^\alpha D_x^\alpha \bar{\ell}(x,\xi), \tag{A.24}$$

then the conjugate T_ℓ^* of T_ℓ is a linear continuous map $H^s \to H^{s-m}$, and $T_\ell^* - T_{\ell^*}$ is a linear continuous map $H^s \to H^{s-m+\rho}$.

Theorem A.22. *Assume $L \in Op\Sigma_\rho^m(\Omega_x)$ with symbol $\ell(x,\xi)$, ψ is a homeomorphism from Ω_x to O_y, then the operator defined by $u \mapsto (L(u\circ\psi))\circ\psi^{-1}$ belongs to $Op\Sigma_\rho^m(O_y)$ with symbol*

$$\sum_\alpha{}' \frac{1}{\alpha!}\ell^{(\alpha)}(x, {}^t\psi'(x)\eta) D_x^\alpha e^{i<\chi(x,z),\eta>}\Big|_{z=x=\psi^{-1}(y)}, \tag{A.25}$$

where $\chi(x,z) = \psi(z)-\psi(x)-\psi'(x)(z-x)$, $\ell^{(\alpha)}(x,\xi)$ means $\partial_\xi^{(\alpha)}\ell(x,\xi)$, and $\sum'$ means the sum of all terms having the form of power of η with degree no less than $m-[\rho]$.

Theorem A.23. *Assume that $F(u_1,\cdots,u_N)$ is a C^∞ function of its arguments. For $1\le j\le N$, $u_j(x)\in C^\rho(R_x^n)$ with $\rho>0$ ($u_j(x)\in H^s(R_x^n)$ with $s>n/2$ resp.). Then*

$$F(u_1(x),\cdots,u_N(x)) = \sum_{j=1}^N T_{\frac{\partial F}{\partial u_j}} u_j(x) + R(x), \tag{A.26}$$

where $R(x)\in C^{2\rho}$ ($R(x)\in H^{2s-n/2}$ resp.). Such an expression of a nonlinear function F is called **paralinearization**.

When F is only finitely regular with respect to its arguments, the corresponding paralinearization still holds, while the regularity of the remainder also depends on the regularity of F.

Theorem A.24. *Assume that $F(u_1,\cdots,u_N)$ is a C^σ ($\sigma>1$) function of its arguments. For $1\le j\le N$, $u_j(x)\in C^{\rho+1}(R_x^n)$ with $\rho>0$. Then*

$$F(u_1(x),\cdots,u_N(x)) = \sum_{j=1}^N T_{\frac{\partial F}{\partial u_j}} u_j(x) + \sum_{k\ge0} F_k\circ(S_k u) + R(x), \tag{A.27}$$

where F_k is the diadic decomposition of F in R^N, S_k is the operator of partial sum introduced in the diadic decomposition in R^n, $R(x)\in C^{\rho+1+\epsilon}$ with $\epsilon=\min(\sigma-1,\rho+1)$.

If $F(u_1,\cdots,u_N)$ is an H^s ($s>N/2+1$) function of its arguments. For $1\le j\le N$, $u_j(x)\in C^{r+1}(R_x^n)$ with $r>n/2$, then Eq. (A.27) still holds with $R(x)\in H^{r+1+\epsilon'}$, $\epsilon'=\min(s-N/2-1, r-n/2+1)$.

By using the paralinearization of nonlinear functions, one can make paralinearization of nonlinear partial differential equations. For a nonlinear partial differential equation

$$F(x, u(x), \cdots, \partial^\beta u, \cdots)_{|\beta|\le m} = 0 \tag{A.28}$$

in R^n, the following theorem holds

Theorem A.25. *Assume that $u \in C^{\rho+m}$, $\rho > 0$ (or $u \in H^{s+m}$, $s > n/2$) is a real solution of Eq. (A.28), then there is a paradifferential operator $P \in Op\Sigma_\rho^m$ with symbol*

$$\sigma(P) = \sum_{|\alpha|\le m} \frac{\partial F}{\partial q_\alpha}(x, u, \cdots, \partial^\beta u, \cdots)(i\xi)^\alpha \in \Sigma_\rho^m, \tag{A.29}$$

such that

$$Pu(x) = \sum_{|\alpha|\le m} T_{\frac{\partial F}{\partial q_\alpha}} \partial^\alpha u(x) \in C^{2\rho} \text{ (or } H^{2s-n/2}). \tag{A.30}$$

A.3 Paracomposition

Next we introduce the concept of paracomposition, which essentially is an extension of the paralinearization of non-smooth functions. Let Ω_{1y}, Ω_{2x} are two given domains, $\chi: \Omega_1 \to \Omega_2$ is a $C^{\rho+1}$ invertible transformation, $\rho > 0$. Let $u \in \mathscr{E}(\Omega_2)$, supp$u \subset K$, then the image of K under the transformation χ^{-1} is $\chi^{-1}(K) \subset \Omega_1$. Assume that $\{\mathscr{C}_j\}, \{\tilde{\mathscr{C}_j}\}$ are two sets of rings introduced in the diadic decomposition of Ω_{2x}, Ω_{1y} respectively. $u(x)$ has a diadic decomposition

$$u(x) = \sum_{j=-1}^{\infty} u_j(x),$$

where $u_j(x)$ satisfies supp$\hat{u}_j \subset \mathscr{C}_j$. Since the variables ξ and η obeys the relation $\eta = \langle {}^t\chi'(y), \xi\rangle$, then one can choose a large number $\tilde{\kappa} > 1$ and a suitable neighborhood $\Omega' \subset\subset \Omega_1$ of $\chi^{-1}(K)$, such that for any $y \in \Omega_1', \xi \in \mathscr{C}_j$, there is $\eta \in \tilde{\mathscr{C}_j} = \{\eta;\ \tilde{\kappa}^{-1}2^j \le |\eta| \le \tilde{\kappa}2^{j+1}\}$.

Definition A.4. Assume that u and χ are given as above, $\psi(y) \in C_0^\infty(\Omega_1)$, and equals 1 on a neighborhood of $\chi^{-1}(K)$. Define

$$\chi^* u = \sum_k [\psi(u_k \circ \chi)]_k. \tag{A.31}$$

The notation $[\ ]_k$ means $[v]_k = \sum_j v_j$, where the index j in the summation runs over the terms, whose spectral intersects with the diadic ring $\mathscr{C}_k$. The operator χ^* is called **paracomposition operator**.

Under the assumption $\chi \in C^{\rho+1}$ with $\rho > 0$, the paracomposition is a linear continuous map from C^σ to C^σ ($\sigma \neq 0$), or from H^s to H^s. In Definition A.4 the different choice of diadic decomposition only cause an error of ρ regular operator.

Theorem A.26. *Assume that* $\chi : \Omega_1 \to \Omega_2$ *is a* $C^{\rho+1}$ (H^{r+1} *resp.*) *invertible transformation,* $\rho > 0$ ($r > n/2$ *resp.*), $u \in C^\sigma(\Omega_2)$, $\sigma > 1$ ($u \in H^s(\Omega_2)$, $s > n/2 + 1$ *resp.*), u *has a compact support in* Ω_2, $\psi \in C_0^\infty(\Omega_1)$ *equals* 1 *in a neighborhood of* supp$u$, *then*

$$u \circ \chi = \chi^* u + T_{u' \circ \chi}(\psi(\chi)) + R, \tag{A.32}$$

where $R \in C^{\rho+1+\epsilon}, \epsilon = \min(\sigma - 1, \rho + 1)$ ($R \in H^{r+1+\epsilon'}, \epsilon' = \min(s - n/2 - 1, r - n/2 + 1)$ *resp.*).

Theorem A.27. *Assume that* $\chi_1 : \Omega_1 \to \Omega_2$, $\chi_2 : \Omega_2 \to \Omega_3$ *are two* $C^{\rho+1}$ *invertible transformation,* $\rho > 0$, $u \in C^\sigma(\Omega_1)$, $\sigma > 1$ ($u \in H^s(\Omega_1)$, $s > n/2 + 1$ *resp.*), u *has a compact support* K *in* Ω_2, $\psi \in C_0^\infty(\Omega_2)$ *equals* 1 *in a neighborhood of* $\chi_2^{-1}(K)$, *then*

$$\chi_1^* \psi \chi_2^* u = (\chi_2 \chi_1)^* u + Ru, \tag{A.33}$$

where R *is a* ρ *regular operator.*

Theorem A.28. *Assume that* $\chi_1 : \Omega_1 \to \Omega_2$ *is a* $C^{\rho+1}$ *invertible transformation,* $\rho > 0$, $u \in C^\sigma(\Omega_2)$, $\sigma > 1$ ($u \in H^s(\Omega_2)$, $s > n/2 + 1$ *resp.*), u *has a compact support* K *in* Ω_2. *Moreover, assume that the paradifferential operator* T_a *has its symbol* $a(x, \xi) \in \Sigma_\alpha^m(\Omega_2)$, *then there is an operator* $T_{\tilde{a}}$, *such that*

$$\chi^* T_a u = T_{\tilde{a}} \chi^* u + Ru, \tag{A.34}$$

where $\tilde{a}(y, \eta) \in \sum_\epsilon^m(\Omega_1)$, $\epsilon = \min(\alpha, \rho)$, *and* R *is a* $\rho - m$ *regular operator.*

Theorem A.29. *Assume that* $\chi_1 : \Omega_1 \to \Omega_2$ *is a* $C^{\rho+1}$ *invertible transformation,* $\rho > 0$ (H^{r+1} *resp.* $r > n/2$), *for* $1 \le j \le N$, $u_j(x) \in C^\sigma(\Omega_2)$, $\sigma > 1$ ($u \in H^s(\Omega_2)$, $s > n/2 + 1$ *resp.*). *Moreover,* $F(u_1, \cdots, u_N)$ *is a* C^∞ *function of its arguments on* R^N, *then*

$$\chi^* F(u) = T_{F'(u \circ \chi)} \chi^* u + R, \tag{A.35}$$

where $R \in C^\epsilon$, $\epsilon = \min(2\rho_2, \rho + \sigma, 2\sigma)$ ($R \in H^\epsilon$, $\epsilon = \min(2r - n/2 + 2, r + s - n/2, 2s - n/2)$ *resp.*).

Bibliography

[1] Albano P. and Cannarsa P. *Propagation of singularities for solutions of nonlinear first order partial differential equations*, Arch. Rat. Mech. Anal., 2002, v.162, pp.1-23.

[2] Alinhac S. *Évolution d'une onde simple pour des équations non-linéaires génerales*, Corent Topics in PDEs, Kinekuniya, Tokyo, 1985, pp.63-90.

[3] Alinhac S. *Paracomposition et operateurs paradifférentiels*, Communications in PDEs, 1986, v.11, pp. 87-121.

[4] Alinhac S. *Interaction d'ondes simples pour des équations complètement non-linéaires*, Ann. Scient. École Norm. Sup., 1988, v.21, pp.91-132.

[5] Alinhac S. *Existence d'ondes de raréfaction pour des systèmes quasi-linéaires hyperbolique multidimensionnels*, Communications in PDEs, 1989, v.14, pp. 173-230.

[6] Ascanelli A. and Cicognani M. *Propagation of singularities in the Cauchy problem for a class of degenerate hyperbolic operators*, Jour. Math. Anal. Appl., 2008, v.341, pp.694-706.

[7] Beals M. *Spreading of singularities for a semilinear wave equation*, Duke Math. J., 1982, v.49, pp.275-286.

[8] Beals M. *Self-spreading and strength of singularities for solutions to semilinear wave equations*, Ann. of Math., 1983, v.118, pp.187-214.

[9] Beals M. *Nonlinear wave equations with data at one point*, Comtemp. Math., 1984, v.27, pp.83-95.

[10] Beals M. *Singularities of conormal radially smooth solutions to nonlinear wave equations*, Communications in PDEs, 1988, v.13, pp.1355-1382.

[11] Beals M. *Vector fields associates with the nonlinear interaction of progressing waves*, Indiana Univ. Math., 1988, v.37, pp.637-666.

[12] Beals R. and Kannai Y. *Exact solutions and branching of singularities for some hyperbolic equations in two variables*, Jour. Diff. Eqs., 2009, v.246, pp.3448-3470.

[13] Beals M. and Metivier G. *Progressing wave solution to certain nonlinear mixed problems*, Duke Math. J., 1986, v.53, pp.125-137.

[14] Beals M. and Metivier G. *Reflection of transversal progressing waves in nonlinear strictly hyperbolic mixed problems*, Amer. J. Math., 1987, v.109,

pp.335-360.

[15] Beals M. and Reed M. *Propagation of singularities for hyperbolic pseudodifferential operators with nonsmooth coefficients*, Comm. Pure Appl. Math., 1982, v.35, pp.169-184.

[16] Beals M. and Reed M. *Microlocal regularity theorems for non-smooth pseododifferential operators and applications to nonlinear problems*, Trans. Amer. Math. Soc., 1984, v.285, pp.159-184.

[17] Ben-Dor G., *Shock Waves Reflection Phenomena (second edition)* Springer-Verlag, Berlin, Heiderberg, New York, 2007.

[18] Bernardi E. *Propagation of singularities for hyperbolic operators with multiple involutive characteristics*, Osaka Jour. Math., 1988, v.25, pp.19-31.

[19] Bony J. M. *Calcul symbolique et propagation des singularités pour les équations aux dériveées nonlinéaires*, Ann. Scient. Ec. Norm. Sup., 1981, v.14, pp.209-246.

[20] Bony J. M. *Propagation des singularités pour les équations aux dériveées partielles nonlinéaires*, Sem. Goulaouic-Schwartz. Ecole Polytechnique, 1979-1980, no.22.

[21] Bony J. M. *Interaction des singularités pour les équations aux dériveées partielles nonlinéaires*, Sem. Goulaouic-Schwartz. Ecole Polytechnique, 1980-1981, no.2.

[22] Bony J. M. *Second microlocalization and propagation of singularities for semilinear hyperbolic equations*, Tanikachi Symp. Katata, 1984, 11-49.

[23] Bony J. M. *Analyse microlocale des équations aux dériveées partielles*, Lecture Notes in Math., 1991, v.1495, pp.1-45.

[24] Chazarian J. *Propagation des singularités pour classe d'opérateurs a caractéristiques multiples et résobulité locale*, Ann. Inst. Fourier, 1974, v.24, pp.203-233.

[25] Chazarian J. *Reflection of C^∞ singularities for a class of operator with multiple characteirstics*, Publ. RIMS Kyoto Univ., 1977, v.12, pp.39-52.

[26] Chemin J.Y. *Interaction de trois ondes dans les équations semilinéaires strictement hyperboliques d'order 2*, Communications in PDEs, 1987, v.11, 1203-1225.

[27] Chemin J.Y. *Interaction contrôleé dans les équations aus dérivées partialles nonlinéires*, Bull. Soc. Math. France, 1988, v.116, pp.341-383 .

[28] Chemin J.Y. *Calcul paradifférentiel précisé et applications à des équations aux dérivées partielles non semilinéaires*, Duke Math. J., 1988, v.56, pp.431-469.

[29] Chemin J.Y. *Évolution d'une singularité punctuelle dans des équations strictement hyperboliques non linéaires*, Amer. J. Math., 1990, v.112, pp.805-860.

[30] Chen S.X. *Pseudodifferential operators with finitely smooth symbols and their applications to quasilinear equations*, Nonlinear Analysis TMA, 1982, v.6, pp.1193-1206.

[31] Chen S.X. *Regularity estimate of solution to semilinear wave equation in higher space dimension*, Science in China, Ser.A, 1984, v.27, pp.924- 935.

[32] Chen S.X. *The reflection and interaction of the singularities of solutions*

to semilinear wave equation in higher space dimension, Nonlinear Analysis TMA, 1984, v.8, pp.1167-1179.

[33] Chen S.X. *Regularity estimate of solutions to semilinear hyperbolic equations in higher space dimension*, Acta Math. Sinica, 1987, v.3, pp.66-76.

[34] Chen S.X. *Smoothness of shock front solutions for system of conservation laws*, Lecture Notes in Math., 1988, no.1306, pp.38-60.

[35] Chen S.X. *Poecewise smooth solutiond of semilinear systems in higher space simension*, Chin. Ann. Math., 1989, v.10(B), pp.361-370.

[36] Chen S.X. *On reflection of multidimensional shock front*, Jour. Diff. Eqs., 1989, v.80, pp.199-236.

[37] Chen S.X. *Reflection and interaction of progressing wave for semilinear wave equations*, Jour. Math. Anal. Appl., 1990, v.153, pp.562-575.

[38] Chen S.X. *Multidimensional Riemann problem for semilinear wave equation*, Communication in PDEs, 1992, v.17, pp.715-736.

[39] Chen S.X. *Existence of local solution to supersonic flow around a three dimensional wing*, Advances in Appl. Math., 1992, v.13, pp.273-304.

[40] Chen S.X. *Pseudodifferential operators.* Advanced Educational Press, 1995(first edition), 2005(second edition), Beijing.

[41] Chen S.X. *M-D Riemann problem for a class of quasilinear hyperbolic system and its perturbation*, Chinese Science Bulletin, 1995, v.40, pp.535-539.

[42] Chen S.X. *On interaction of shock and sound wave (I),(II)*, Chin. Ann. Math., 1996, v.17(B), pp.35-42,445-456.

[43] Chen S.X. *Construction of solutions to M-D Riemann Problems for a 2X2 quasilinear hyperbolic system*, Chin. Ann. Math., 1997, v.18B, pp.345-358.

[44] Chen S.X. *Formation-Construction of shock in compressive simple wave of NXN hyperbolic system*, Proceedings of the Conference Partial Differential Equations and Their Applications. World Scientific, 1999, pp.32-43,

[45] Chen S.X. *Existence of stationary supersonic flow past a pointed body*, Arch. Rat. Mech. Anal., 2001, v.156, pp.141-181.

[46] Chen S.X. *How does a shock in supersonic flow grows out of smooth data*, Jour. Math. Phys., 2001, v.42, pp.1154-1162.

[47] Chen S.X. *Stability of a Mach configuration*, Comm. Pure Appl. Math. 2006, v.59, 1-35.

[48] Chen S.X. *Mach configuration in pseudo-stationary compressible flow*, Jour. Amer. Math. Soc., 2008, v.21, 63-100.

[49] Chen S.X. and Dong L.M. *Formation of shock in potential flow*, Advances in Nonlinear PDEs and Related Areas, World Scientific, 1998, pp.45-66.

[50] Chen S.X. and Dong L.M. *Formation of shock for p-system with general smooth initial data*, Science in China, 2001, v.44, pp.1154-1162.

[51] Chen S.X., Li C.Z. and Qiu Q.J. *Introduction to paradifferential operators*, Science Press, Beijing, 1990.

[52] Chen S.X. and Wang Y.G. *The Cauchy problem for gas dynamic systems in multi-dimensional space with weakly singular data*, Chin. Ann. Math., 1992, v.13(B), pp.298- 314.

[53] Chen S.X. and Wang Y.G. *Propagation of singularities in compressible viscous fluids*, *AMS/IP Stud. Adv. Math.*, 2002, v.29, pp.13-25.

[54] Chen S.X. and Wang Y.G. *Propagation of singularities of solutions to hyperbolic-parabolic coupled system*, Mathematishe Nachschrift, 2002, v.242, pp.46-60.

[55] Chen S.X., Xin Z.P. and Ying H.C. *Global sock waves for the supersonic flow past a perturbed cone*, Comm. Math. Phys., 2002, v.228, pp.47-84.

[56] Chen S.X. and Zhang Z.B. *On the generation of shock waves for first order quasilinear equation*, Fudan Journal (Natural Sciences), 1963, v.1,13-22.

[57] Chen G.Q. and Feldman M., *Potential theory for shock reflection by a large-angle wedge.* Proc. Nat. Acad. Sci. USA., 2005, v.102, pp.15368-15372. *Global solution to shock reflection by large-angle wedges for potential flow.* Ann. Math.,accepted 2006.

[58] Coulombel J.F. and Secchi P., *Nonlinear compressible vortex sheets in two space dimensions*, Ann. Sci. Ec. Norm. Super.,2008, v.41, pp.85-139.

[59] Courant R. and Friedrichs K.O. *Supersonic flow and shock waves*, Springer-Verlag, New York, 1976.

[60] David F. and Williams M. *Singularities of solutions to semilinear boundary value problems*, Amer. Jour. Math., 1987, v.109, pp.1087-1109.

[61] Dencker M. *On the propagation of singularities for pseudo-differential operators with characteristics of variable multiplicity*, Communications in PDEs, 1992, v.17, pp.1709-1736.

[62] Dreher M. *Weakly hyperbolic equations, Sobolev spaves of veriable order, and propagation of singularities*, Osaka Jour. Math., 2002, v.39, pp.409-445.

[63] Dreher M. and Reissig M. *Propagation of mild singularities for semilinear weakly hyperbolic equations*, Jour. Anal. Math., 2000, v.82, pp.233-266.

[64] Duan R.J., Li M.R. and Yang T. *Propagation of singularities in the solutions to the Boltzmann equation near equibrium*, Math. Models Mathods Appl. Sci., 2008, v.18, pp.1093-1114.

[65] Garåding L. *Singularities in linear wave propagation*, Lecture Notes in Math., 1987, no.1241.

[66] Gerard P. *Solutions conormales analytique d'équations hyperbolique non-linéaires.*, Communications in PDEs, 1988, v.13, pp.345-375.

[67] Gerard P. and Rauch J. *Propagation de la régularité locale de solutions d'équations hyperboliques nonlinéaries*, Ann. Inst. Fourier, 1987, v.37, pp.65-84.

[68] Godin P. *Propagation of C^∞ regularity for fully nonlinear second order strictly hyperbolic equations in two variables*, Trans. Amer. Math. Soc., 1985, v.290, pp.825-830.

[69] Hanges N. *Propagation of singularities for a class of operators with double characteristics*, Annals Math. Stud., 1979, v.91, pp.113-126.

[70] Hoff D. and Santos M. *Lagrangean structure and propagation of singularities in multidimensional compressible flow*, Arch. Rat. Mech. Anal., 2008, v.188, pp.509-543.

[71] Hömander L. *On the existence and the regularity of solutions of linear pseudodifferential equations*, Ens. Math. 1971, v.17, pp.99-163.

[72] Hömander L. *Fourier Integral Operators I.* Acta Math., 1971, v.127, pp.79-

183.

[73] Hörmander L. *Spectral analysis of singularities*, Ann. Math. Study, Prieston University Press, 1979, v.91, pp.3-50.

[74] Hörmander L. *The analysis of linear partial differential equations (I)-(IV)*, Springer-Verlag, Berlin, Heidelberg, New York, 1985.

[75] Hörmander L. *Lectures on nonlinear hyperbolic differential equations*, Springer-Verlag, 1997, Berlin, Heiberberg, New York.

[76] Ito K. *Propagation of singularities for Schrödinger equations on the Euclidean space with a scattering metric*, Communications in PDEs, 2006, v.31, pp.1735-1777.

[77] Ito S. *Propagation of singularities for semi-linear wave equations with nonlinearity satisfying the null condition*, Jour. Hyper. Diff. Eqs., 2007, v.4, pp.197-205.

[78] Izumiya S. *Singularites of solutions for first order partial differential equations*, London Math. Soc. Lecture Note, 1999, Ser.263, pp.419-440.

[79] Joly J.L., Metivier G. and Rauch J. *Diffractive nonlinear geometric optics with rectification* Indiana Univ. Math. J., 1998, v.47, pp.1167-1241.

[80] Kohn J.J. and Nirenberg L. *On the algebra of pseudo-differengtial operators*, Comm. Pure Appl. Math., 1965,v.18,pp.269-305.

[81] Kong D.X. *Formation and propagation of singularities for* 2×2 *quasilinear hyperbolic systems*, Trans. Amer. Math. Soc., 2002, v.354, pp.3155-3179.

[82] Lascar B. *Singularités des solutions d'équations aux dérivées partielles non linéaires*, CRAS Paris, 1978, v.287, pp.527-529.

[83] Lax P.D. *Asymptotic solutions of oscillatory initial value problems*, Duke Math. J., 1957, v.24, pp.627-646.

[84] Lax P.D. *Hyperbolic systems of conservation laws and the mathematical theory of shock waves*, Conf. Board Math. Sci., 11, SIAM,(1973).

[85] Lebeau G. *Équations des ondes semi-linéaries II. Contrôle des singularités et caustiques non linéaires*, Invent. Math., 1989, v.95, pp.277-323.

[86] Li D.N. *Rerefaction and shock waves for multidimensional hyperbolic conservation laws*, Communications in PDEs., 1991, v.16, pp.425-450.

[87] Li D.Q. and Chen S.X. *Regularity and singularity of solutions for nonlinear hyperbolic equations*, Advances in Science of China, 1991, v.1, pp.129-162.

[88] Li T.T. and Yu W.C. *Boundary value problems for hyperbolic systems*, Duke Univ. Math. Ser.5, 1985.

[89] Liu L.Q. *Propagation of singularities for semilinear wave equation*, Canad. Jour. Math., 1993. v.45, pp.835-846.

[90] Lien W.C. and Liu T.P., *Nonlinear stability of a self-similar 3-D gas flow*, Comm. Math. Phys., 1999, v.304, pp.524-549.

[91] Li J.Q. and Zheng Y., *Interaction of rarefaction waves of the two-dimensional self-similar Euler equations*, Arch. Rat. Mech. Anal., 2009, v.193, pp.623-657.

[92] Majda A. *The existence of multi-dimensional shock fronts*, Memoirs AMS, 1983, v.283.

[93] Majda A. and Thomann E., *Multidimensional shock fronts for second wave equations*, Communications in PDEs, 1987, v.12, pp.777-828.

[94] Melrose R. *Microlocal parametrices for diffractive boundary value problems*, Duke Math. J., 1975, v.42, pp.605-635.

[95] Melrose R. *Conormal rings and semilinear wave equations*, Advances in Microlocal Analysis, 1986, pp.225-251.

[96] Melrose R. and Ritter N. *Interaction of progressing waves for semilinear wave equation I*, Ann. of Math., 1985, v.121, pp.187-213.

[97] Melrose R. and Ritter N. *Interaction of progressing waves for semilinear wave equation II*, Arkiv for Math., 1987, v.25, pp.91-114.

[98] Melrose R., Sá Barreto A. and Zworski M. *Semi-linear diffraction of conormal waves*, Asterisque, 1996, no. 240.

[99] Melrose R. and Sjöstrand J. *Singularities of boundary value problems I,II*, Comm. Pure Appl. Math., 1978,v.31,pp.593-617, 1982, v.35, pp.129-168.

[100] Melrose R., Vasy A. and Wunsch J. *Propagation of singularities for the wave equation on edge maniforlds*, Duke Math. J., 2008, v.144, pp.109-193.

[101] Melrose R. and Wunsch J. *Propagation of singularities for the wave equation on conic manifolds*, Invent. Math., 2004, v.156, pp.235-299.

[102] Metivier G., *The Cauchy problem for semilinear hyperbolic systems with discontinuous data*, Duke Math. J., 1986, v.53, pp.983-1011.

[103] Metivier G., *Interaction de deux chocs pour un systeme de deux lois conservation en dimension deux d'espace*, Trans. Amer. Math. Soc., 1986, v.296, pp.431-479.

[104] Metivier G., *Propagation, interaction and reflection of discontinuous progressing waves for semilinear hyperbolic systems*, Amer. Jour. Math., 1989, v.111, pp.239-287.

[105] Metivier G., *Ondes soniqes*, Jour. Math. Pures et Appl., 1991, v.70, pp.197-268.

[106] Metivier G. and Rauch J. *The interaction of two progressing waves*, Lecture Notes in Math., 1989, v.1402, pp.216-226.

[107] Mnif M., *Probleme de Riemann pour une loi conservation scalaire hyperbolique d'order deux*, Communications in PDEs, 1997, v.22, pp.1589-1627.

[108] Nirenberg L. *Lecture on linear partial differential equations*, Amer. Math. Soc., Reginal Conf. in Math., 1972, v.17, pp.1-58.

[109] Payne K. R. *Propagation of singularities phenomena for equations of Tricomi type*, Appl. Anal., 1998, v.68, pp.195-206.

[110] Petrini M. and Sordoni V. *Propagation of singularities for hyperbolic operators with multiple involutive characteristics*, Osaka Jour. Math., 1991, v.28, pp.911-933.

[111] Petrini M. and Sordoni V. *Propagation of singularities for hyperbolic operators with double characteristics*, Ann. Mat. Pura Appl., 1993, v.163, pp.199-222.

[112] Prasad P. *Formation and propagation of singularities on a nonlinear wavefront and a shock front*, Jour. Indian Inst. Sci., 1995, v.75, pp.537-558.

[113] Qi M.Y. *Propagation of singularities for solutions to nonlinear differential equations with constant multiple characteristics*, Jour. Math. Anal. Appl., 1993, v.173, pp.557-576.

[114] Rack R. and Wang Y.G. *Propagation of singualrities in one-dimensional*

thermoelasticity, Jour. Math. Anal. Appl. 1998, v.223, pp.216-247.

[115] Rauch J. *Singularities of solutions to semilinear wave equations*, Jour. Math. Pures et Appl., 1979, v.58, pp.299-308.

[116] Rauch J. and Reed M. *Jump discontinuities of semilinear, strictly hyperbolic systems in two variables; creation and propagation*, Comm. Math. Phys., 1981, v.81, pp.203-227.

[117] Rauch J. and Reed M. *Singularities produced by the nonlinear interaction of three progressing waves: example*, Communications in PDEs, 1982, v.7, pp.1117-1133.

[118] Rauch J. and Reed M. *Nonlinear microlocal analysis of semilinear hyperbolic systems in one space dimension*, Duke Math. J., 1982, v.49, pp.397-475.

[119] Rauch J. and Reed M. *Discontinuous progressing waves for semilinear systems*, Communications in PDEs, 1985, v.10, pp.1033-1075.

[120] Rauch J. and Reed M. *Straited solutions of semilinear two speed wave equations*, Indiana Math. Jour., 1985, v.34, pp.337-353.

[121] Rauch J. and Reed M. *Classical conormal solutions of semilinear systems*, Communications in PDEs, 1988, v.13, pp.1297-1335.

[122] Sablé-Tougeron M. *Régularité microlocale pour des problèmes aux limites non-linéaires*, Ann. Inst. Fourier, 1986, v.36, pp.39-82.

[123] Sablé-Tougeron M. *Ondes de gradients multidimensionnelles*, Memoirs AMS, 1993, no.511.

[124] Sà Barreto A. *Interaction of conormal waves for fully semilinear wave eqaution*, Jour. Func. Anal., 1990, v.89, pp.233-273.

[125] Sà Barreto A. *Second microlocal ellipticity and propagation of conormality for semilinear wave equations*, Jour. Func. Anal., 1991, v.102, pp.47-71.

[126] Sideris T. *Formation of singualrities in solutions to nonlinear hyperbolic equations*, Arch. Rat. Mech. Anal., 1984, v.86, pp.369-381.

[127] Sjöstrand J. *Singularités analytiques microlocales*, Astéristique, 1982, no.95.

[128] Smoller J. *Shock waves and reaction-diffusion equations*, Springer-Verlag, New York, Heiderberg, Berlin, 1982.

[129] Suchkov V.A. *Flow into a vacuum along an oblique wall*, Jour. Appl. Math. Mech., 1963, v.27, pp.1132-1134.

[130] Taylor M. *Refection of singularities of solutions to systems of differential equations*, Comm. Pure Appl. Math., 1975, v.28, pp.457-478.

[131] Taylor M. *Grazing rays and refection of singularities to wave equations*, Comm. Pure Appl. Math., 1976, v.29, pp.1-38.

[132] Taylor M. *Pseudodifferential operators*, Prinston University Press, Prinston,1981.

[133] Tricomi, F. *Sulle equazioni lineari alle derivate parziali di secondo ordino, di tipo misto*, Rendiconli, Alli dell' Accademia Nazionalw dei Lincei, 1923, v.14, pp.134-247.

[134] Vasy A. *Propagation of singualrities in many-body scattering in the presence of bound states*, Jour. Funct. Anal., 2001, v.184, pp.177-272.

[135] Vasy A. *Propagation of singualrities in many-body scattering*, Ann. Sci.

École Norm. Sup., 2001, v.34, pp.313-402.

[136] Vasy A. *Propagation of singularities for the wave equation on manifolds with conners*, Ann. of Math., 2008, v.168, pp.749-812.

[137] Wang W.K. *Interaction of shock and conormal singularities for conservation laws*, Communications in PDEs, 1994, v.19, pp.1037-1073.

[138] Wang Y.G. *The semilinear second-order hyperbolic equation with data strongly singular at one point*, Monatshefte für Math., 1994, v.117, pp.285-302.

[139] Wang Z.J. and Zhang Y.Q. *Steady supersonic flow past a curved cone* Jour. Diff. Eqs., 2009, v.247, pp.1817-1850.

[140] Williams M. *Spreading of singularities at the boundary in semilinear hyperbolic mixed problems I: microlocal $H^{s,s'}$ regularity*, Duke Math, J., 1988, v.56, pp.17-40.

[141] Williams M. *Spreading of singularities at the boundary in semilinear hyperbolic mixed problems II: crossing and self-spreading*, Trans. Amer. Math. Soc., 1989, v.311, pp.291-321.

[142] Williams M. *Highly oscillatory multidimensional shock*, Comm. Pure Appl. Math., 1999, v.52, pp.129-192.

[143] Wunsch J. *Propagation of singularities and growth for Schrödinger operators*, Duke Math. J., 1999, v.98, pp.137-186.

[144] Xu C.J. *Régularité des solutions d'équations aux dérivées partilles nonlinéaires associées à un système de champs de vecteurs*, Ann. Inst. Fourier, 1987, v.37, pp.105-113.

[145] Xu C.J. *Propagation au bord des singularités pour des problèmes de Dirichlet nonlinéaires d'order deux*, Jour. Func. Anal., 1990, v.92, pp.325-347.

[146] Yamamoto K. *Reflective and refractive phenomena of tangential elastic waves to the boundary in two solids as a propagation of singularities*, Jour. Math. Pures Appl., 2009, v.92, pp.188-208.

[147] Yin H.C. and Qiu Q.J. *Tangent interaction of cornomal waves for second order full nonlinear strictly hyperbolic equations*, Nonlinear Anal., 1992, v.19, pp.81-93.

[148] Zhang Y.Q., *Steady supersonic flow past an almost stright wedge with large vertex angle*, Jour. Diff. Eqs, 2003, v.192, pp.1-46.

[149] Zworski M. *An example of new singularities in the semilinear interaction of a cusp and a plane*, Communications in PDEs, 1996, v.21, pp.901-909.